Physics of Intense Beams in Plasmas

T0203947

Plasma Physics Series

Other books in the series

An Introduction to Alfven Waves
R Cross

MHD and Microinstabilities in Confined Plasma
W M Manheimer and C N Lashmore-Davies

Transition Radiation and Transition Scattering
V L Ginzburg and V N Tsytovich

Plasma Diagnostics based on Forward Angle Scattering
L Sharp, J Howard and R Nazikian

Radiofrequency Heating of Plasmas
R A Cairns

Plasma Physics via Computer Simulation
C K Birdsall and A B Langdon

Electromagnetic Instabilities in an Inhomogeneous Plasma
A B Mikhailovskii

Tokamak Plasma: A Complex Physical System
B B Kadomtsev

Plasma Physics Series

Physics of Intense Beams in Plasmas

Mikhail V Nezlin

Russian Scientific Center, Kurchatov Institute,
Moscow, Russia

Translation Editor: Professor E W Laing

Translated from the Russian by Vitaly I Kisin

CRC Press
Taylor & Francis Group
Boca Raton London New York

CRC Press is an imprint of the
Taylor & Francis Group, an **informa** business

CRC Press
Taylor & Francis Group
6000 Broken Sound Parkway NW, Suite 300
Boca Raton, FL 33487-2742

First issued in paperback 2019

© 1993 by Taylor & Francis Group, LLC
CRC Press is an imprint of Taylor & Francis Group, an Informa business
Typeset in TEX at IOP Publishing Ltd

No claim to original U.S. Government works

ISBN-13: 978-0-367-40249-5

Visit the Taylor & Francis Web site at
http://www.taylorandfrancis.com

and the CRC Press Web site at
http://www.crcpress.com

British Library Cataloguing in Publication Data.
A catalogue record for this book is available from the British Library.

Library of Congress Cataloging-in-Publication Data are available

Contents

Preface   ix

Acknowledgments   xi

1   Introduction and General Overview   1

    References   8

2   Aperiodic Beam Instabilities (Theory)   10

  2.1   Bursian Instability and Limiting Current of a Beam of
       Particles of Identical Sign in Vacuum   10
  2.2   Instability and Limiting Current of Quasineutral
       (Compensated) Electron Beams   14
  2.3   Instability and Limiting Current of Quasineutral Ion
       Beams   22
  2.4   Instability and Limiting Current of Relativistic Electron
       Beams   23
       References   24

3   Oscillatory Beam Instabilities (Theory)   26

  3.1   Oscillatory Pierce Instability and its Evolution to
       Chaos: from Regular Oscillations via a Sequence of
       Period-Doubling Bifurcations to a Stochastic Attractor and
       its Crisis   26
  3.2   Electron–Ion Instability of Quasineutral Electron Beams   30
  3.3   Electron–Electron Instability   38
  3.4   Beam–Drift Instability of a Spatially Non-uniform
       Beam–Plasma System in a Magnetic Field   40
  3.5   Thresholds of Electron–Ion Instabilities of Relativistic
       Electron Beams   44
  3.6   On Slipping-Stream Instabilities   46
       References   47

4   Beam Instabilities as a Result of Active Coupling between Waves of Different Signs of Energy (Theory)   49

4.1   Waves of Positive and Negative Energy in Charged Particle Beams   49

4.2   Instability in Interaction Between Waves with Opposite Signs of Energy   54

4.3   Elementary Processes at the Foundation of Beam Instabilities   56

4.4   Analogy of the Induced Anomalous Doppler Effect to the Instability of a Negative-Energy Wave   58

4.5   Effect of the Spread of Electron Beam Velocities on the Type of Beam Instability   59

4.6   Effects of Dissipation and Collisions on Beam Instability in Plasmas   62

4.7   On Negative-Energy Waves in Hydrodynamics   64

        References   65

5   Electron, Ion and Plasma Beams in Laboratory Plasma   67

5.1   Neutralization of Electron Beam Space Charge, Quasineutral Beams, Plasma Beams   67

5.2   Beam–Plasma Discharge   69

5.3   On Temperature of Particles in Beam Plasma   70

5.4   Hot-Cathode Discharge as a Means of Producing Plasma Beams   72

5.5   High-Power Pulsed Ion Beams   77

5.6   Measurement of Space Potentials, Electric Fields and Particle Energies and Concentration in Beam Plasma   78

        References   90

6   Experimental Data on Instabilities and Limiting Currents of Electron, Ion and Plasma Beams. Mechanisms of Current Limitation (Disruption) in Beams   94

6.1   The Limiting Electron Beam Current in the Absence of Space Charge Neutralization   94

6.2   The Limiting Current of Quasineutral Electron Beam   97

6.3   Instabilities Responsible for Current Limitation (Disruption) in Quasineutral Electron Beams: Beam–Drift and Pierce Instabilities   109

6.4   Budker–Buneman Electron–Ion Instability. Does it Produce Current Disruption?   118

6.5   The Electron–Electron Instability   124

6.6 Plasma Stabilization (Raising of Thresholds) of
     Electron–Ion Beam Instabilities and of the Pierce
     Instability   132
6.7 Limiting Currents of Ion and Relativistic Electron Beams 135
6.8 On the Efficiency of Application of Quasiclassical
     Approximation to an Analysis of Large-Scale Beam
     Instabilities   137
6.9 Instability of Plasma Beams. Formation of the Virtual
     Cathode in Plasma Beams   137
6.10 Experimental Observation of Diocotron Instability   141
     References   142

**7 Acceleration and Heating of Plasma Ions in a Regime
    Close to the Limiting Beam Current   145**

7.1 Heating of Ions in Unstable Plasma Beams. An Unstable
     Plasma Beam as an Injector of Hot Ions into a Plasma
     Trap with Magnetic Mirrors   145
7.2 Acceleration of Ions in Plasma Beams with Moving Virtual
     Cathode   151
     References   156

**8 Instability of Intense Ion Beam with Neutralized Space
    Charge in Magnetic Field   158**

8.1 About this Chapter   158
8.2 Mechanism of Oscillation Amplification in Ion Beams   160
8.3 Similarity Laws for Plasma Ion Sources   168
8.4 Methods for Drastically Improving the Neutralization of
     the Ion Beam Space Charge   169
     References   171

**9 Electric Double Layers in Electron, Ion and Plasma
    Beams and the Mechanism of Aurora Borealis   172**

9.1 Double Layers in the Plasma (Theory)   172
9.2 Double Layers in the Laboratory. A Virtual Cathode in an
     Electron Beam as a Double Layer   174
9.3 Double Electric Layer ('Virtual Anode') in an Ion Beam
     Propagating in a Plasma, in Earth's Magnetic Dipole
     Model. The Mechanism of Polar Auroras   176
     References   183

10 Langmuir Solitons in Electron (Plasma) Beams   186

10.1 Solitons in Beam Physics   186

10.2 Physical Concept of Self-Compression (Collapse) of
      Langmuir Waves and Soliton Formation   189
10.3 Langmuir Cavitons in Non-Magnetized Plasma
      (Experiment)   201
10.4 Langmuir Solitons in Magnetized Plasma   208
10.5 Oblique Langmuir Solitons in Magnetized Plasma   228
10.6 Fast Solitons in Magnetized Plasma Waveguides   237
10.7 Upper-Hybrid Solitons   238
10.8 On Ionization-Wave Solitons in Gas Discharge   242
10.9 On Langmuir Solitons in Outer Space   242
      References   244

11 Novel Methods of Generation and Amplification of
    Electromagnetic Waves by Relativistic Electron Beams   250

11.1 An Electron Beam at a Current above Limiting Bursian
      Current as a Vircator (Turbotron)   250
11.2 Compton Scattering of Photons by Relativistic Electrons
      (Theory)   253
11.3 Raman Scattering of Electromagnetic Waves by
      Negative-Energy Waves of Electron Beams (Theory)   259
11.4 Free Electron Lasers (Masers) Using Compton and Raman
      Scattering (Experiment)   265
      References   271

Conclusion   275

Index   276

Preface

This volume is devoted to the physics of collective phenomena in beams of charged particles which propagate through plasma; it also describes physically related phenomena in intense beams propagating through the vacuum.

The following factors determine the fundamental physical importance of the subjects treated in this book. First, this is an unalienable part of plasma physics. Historically, it was the study of the propagation of electron beams in plasmas that led the founder of the science of plasma, Irving Langmuir, to discovering the first collective phenomena in plasma, such as the generation of plasma oscillations (Langmuir oscillations) and the formation of electric double layers. Later, the collective phenomena in beams travelling through plasma invariably served as a 'proving ground' for the elaboration of plasma theory including such notions as Landau damping (direct and inverse), Čerenkov and Doppler (normal and anomalous) collective effects and a number of plasma instabilities. It was in beam physics that the concept of negative-energy waves had appeared in the 1950s; this concept proved to be an exceptionally fruitful approach to wave dynamics in physical electronics, radiophysics, hydrodynamics, plasma physics and solid state physics. Relatively recently, it was in electron beams propagating through plasma that a new fundamental object of non-linear physics—the Langmuir soliton—and such general non-linear wave phenomena as Langmuir wave collapse were discovered. Furthermore, natural phenomena, such as the aurora borealis and intense radio emission in planetary atmospheres, also involve beam physics.

Second, the utilization of intense electron beams is the basis for designing the most powerful and easily retunable generators of widest-range oscillations, among them the so-called *free electron lasers and masers* and high-power charged particle accelerators. High-intensity ion beams are efficiently applied to the heating of plasmas and maintaining the current in modern magnetic traps for fusion plasma, to electromagnetic isotope separation etc.

Obviously, collective phenomena manifest themselves especially clearly in high-intensity beams. It is therefore beyond doubt that a comprehensive analysis of the physics of the propagation of intense beams through plasmas is very desirable. It is quite strange, however, that monographs covering this field are typically neither complete nor sufficiently comprehensive. Thus some of them treat exclusively the theories relevant to the field (such as M Kuzelev and A Rukhadze 1990 *Electrodynamics of Dense Electron Beams in Plasma*), while other monographs are mostly concerned with the physics of relativistic electron beams (such as R B Miller 1982 *An Introduction to the Physics of Intense Charged Particle Beams*). For these reasons, I intended to write a sufficiently comprehensive review of the collective phenomena in beams which would not suffer from the above restrictions.

An attempt to write the review led to this book, which considers a consistent series of experimental studies devoted to investigating the most important collective phenomena (instabilities) in beams of all energies (including relativistic beams). A theoretical approach, which is unlikely to create difficulties for the reader, is outlined in detail; it allows us to analyse the collective phenomena efficiently. It is important that neither the wide scope of phenomena nor the generality of the approach to their analysis prevent the book from being quite original: it is built to a large extent on the research of my colleagues and myself carried out at the I V Kurchatov Institute of Atomic Energy (now the Russian Scientific Center, Kurchatov Institute).

Part of the material presented in this book appeared in the Russian edition a decade ago (M V Nezlin 1982 *Dynamics of Beams in Plasmas* (Moscow: Energoizdat)). The present volume, specially written for the readership in the West, covers a considerably wider scope of problems and gives an updated and extended analysis of the phenomena described in the 1982 edition. The book is mostly aimed at physicists engaged in plasma research, non-linear physics, physical electronics and radiophysics.

It is my pleasant obligation to express the deep gratitude to all colleagues whose help and advice assisted me in conducting the experiments and in analysing their results.   My special thanks go to my closest co-workers A M Solntsev, A S Trubnikov, M I Taktakishvili, E N Snezhkin, S V Antipov and to those colleagues who took part in numerous discussions and seminars: Professors M A Leontovich, B B Kadomtsev, M S Ioffe, Ya B Fainberg, A B Mikhailovsky, A A Rukhadze, V V Vladimirov, A V Zharinov and V V Arsenin.

M V Nezlin
August 1992, Moscow

Acknowledgments

The author is grateful to the following for granting permission to reproduce figures included in this book.

American Physical Society for figures 6.39, 7.9, 9.4, 10.1, 10.2, 10.3, 10.4

The American Institute of Physics for figures 3.4, 9.6, 10.5, 11.5

The American Geophysical Union for figure 10.31

The Institute of Electrical and Electronics Engineers, Inc.

Davis H A *et al* 1988 *IEEE Trans.: PS*-16 192 (figure 11.1 ©1988 IEEE)

Birkett D S *et al* 1981 *IEEE Trans.: QE*-17 1348 (figures 11.2, 11.6, 11.7 ©1981 IEEE)

The author and IOP Publishing Ltd have attempted to trace the copyright holder of all the figures reproduced in this publication and apologize to copyright holders if permission to publish in this form has not been obtained.

1 Introduction and General Overview

In this book, we discuss collective (wave) interactions in beams of charged particles, which propagate in plasma and in plasma-like media; we also consider the main consequences of these interactions, namely, beam instabilities. This subdivision of plasma physics, discovered in the fundamental work of Langmuir and Tonks [1.1–1.3], Pierce [1.4–1.7], Haeff [1.8, 1.9], Akhiezer and Fainberg [1.10–1.13] and Bohm and Gross [1.14], has been comprehensively developed in the last four decades. The physics of plasma beams attracts researchers mostly for two reasons. First, a quasineutral beam of charged particles (either in a plasma or taken independently of the medium) is the most instructive example of the simplest plasma system far from the state of thermodynamic equilibrium. As a result, quasineutral beams reveal various plasma instabilities in the most clear-cut and diversified manner. The specifics of the velocity distribution function of particles (in a monoenergetic beam) produce a new quality, that is, instabilities gain features which are not found in 'ordinary' plasma (with approximately Maxwellian distribution of particles in the velocity space); consequently, the chances to study plasma instabilities are better in principle and, especially important, the possibilities of experimental testing are significantly improved. One example is the appearance of the threshold (critical) current of the instabilities of a spatially non-uniform quasineutral beam; this threshold depends in a very informative manner on the parameters of the beam, system geometry, external magnetic field, and so on. Neutralized beams of charged particles are thus a very convenient 'proving ground' for plasma theory. This book discusses experimental studies of such beams in substantial detail. Second, intense beams of charged particles are used in a number of applications: heating of plasma in experimental studies of controlled nuclear fusion, new methods of particle acceleration to high energies, plasma chemistry, space research and space technology and, of course, in physical electronics— in high-power all-wave generators of electromagnetic waves. A special

1

place in this field has been taken in recent years by lasers and masers based on collective scattering of electromagnetic waves by high-intensity electron beams (the so-called free electron lasers), by generators (known as vircators) with a virtual cathode produced in an electron beam with high space charge owing to a beam instability, and by cyclotron autoresonance masers. The book pays considerable attention to the principles of operation of these generators.

The following features single out the present book among a number of other books on plasma physics and on the physics of plasma beams.

(1) The book is mostly devoted to experimental studies.

(2) The object of study is high-intensity (high-density) beams. The criterion of high beam intensity can be written in two equivalent forms. The first of them corresponds to the condition that the beam is a plasma system, that is, that the Debye radius of its particles (r_{D1}) is smaller than the beam size (length L, diameter d). In the non-relativistic case, this criterion is written as

$$L, \, d > r_{D1} = (mu_0^2/4\pi n_1 e^2)^{1/2} \tag{1.1}$$

where n_1 and u_0 are the concentration and velocity of particles of the beam; m and e are their mass and charge, respectively. The second form of the high-intensity beam criterion arises because, as we see from (1.1), this beam carries a current (I) exceeding the limiting current of particles of identical sign in vacuum (I_0); this limiting current is proportional to u_0^3 and to the relativistic factor γ_0:

$$I > I_0 \sim \gamma_0 u_0^3 \tag{1.2}$$

where $\gamma_0 = (1 - u_0^2/c^2)^{-1/2}$, and c is the velocity of light. Equations (1.1), (1.2) show that the concept of a high-intensity beam, applied here, is definitely not equivalent to the concept of a high-power beam. For instance, criteria (1.1), (1.2) may be violated even by a very high-power beam if it is relativistic, while a relatively low-power non-relativistic beam with a sufficiently high 'perveance' I/u_0^3 satisfies them.

(3) Beams satisfying the criteria (1.1) and (1.2) of high beam intensity require for their propagation that the space charge be neutralized by particles of the opposite sign. However, a neutralized beam also undergoes instabilities owing to a strong thermodynamic non-equilibrium. Some of these instabilities are so strong that they may cause blocking of beam propagation. The main subjects of this book are precisely the strongest beam instabilities and also their most important corollaries.

If the instability 'bares' a large space charge in the electron beam, which stops the propagation of its electrons, we say that a 'virtual cathode' has been formed in the electron beam—the double electric layer which reflects a considerable proportion of the beam particles (the corresponding phenomenon in an ion beam is known as the 'virtual anode'). If the system contains at the same time, slow plasma particles, they can now be accelerated to fairly high energies by the electric fields of the double layer (created at the expense of the energy of fast particles).

It has recently been shown that in all likelihood, very similar non-linear effects result in one of the mechanisms of auroral phenomena in the atmosphere of the Earth. Chapter 9 of this book presents the modern concept according to which a double electric layer is formed in the plasma containing the ion flow from the magnetosphere of the Earth into its ionosphere; the ions forming this double layer are reflected from it. The potential difference in the layer whose positive pole faces the ionosphere accelerates the electrons arriving from the magnetosphere up to energies in the neighbourhood of the energy of magnetospheric ions (several keV); this is why these electrons can cause the aurora borealis. The physics discussed in this part of the book is closely related with those sections which treat the phenomena involved in the formation of the virtual cathode (or virtual anode) in beams of charged particles.

(4) The general feature of these—the strongest—beam instabilities is their hydrodynamic nature which manifests itself in the fact that all beam particles act coherently; this behaviour is in contrast with the kinetic instabilities that are typically treated in plasma physics books. The latter phenomena are caused by only a small group of 'resonant' particles and are based on the effect of inverse Landau damping. The kinetic instabilities are caused by non-coherent motion of charged particles and are therefore relatively weak (they only widen the energy distribution function of the beam particles), while some hydrodynamic instabilities produce the disruption of the beam current. In contrast to kinetic instabilities, the hydrodynamic beam instabilities arise above a certain critical current (instability threshold) which is found to be quite close to the limiting beam current. Most books on plasma theory pay very little attention to instability thresholds because beam instabilities are typically treated as kinetic effects—opposites of Landau damping (see, e.g., [1.15, 1.16]). In this monograph, however, the problem of beam instability thresholds is of prime importance.

(5) The main hydrodynamic instabilities of a neutralized beam are: beam drift instability, Pierce instability and Budker–Buneman instability.

The first of these, the *beam-drift instability*, is a modification of the drift (or 'universal') instability of magnetized plasma, which is the instability that seems to be a factor of principal importance for the possibility of effective confinement of hot plasma in magnetic traps. When the electron distribution function is not Maxwellian but beam-like, the instability begins at a certain critical beam current. Rich material collected on the basis of the dependence of this critical current on magnetic field, on the energy of the beam electrons and beam size, has made it possible to study in great detail the physics of instabilities and also to stimulate the theory and to recast it in adequate form.

The *Pierce instability*, under certain conditions, is an aperiodic instability which also leads to a disruption of the beam current. The threshold of this instability of the quasineutral beam is higher than the threshold of the beam-drift instability in a relatively weak magnetic field, and is roughly equal to this threshold in strong magnetic fields. As far as physics is concerned, this instability is not very different from the Bursian instability, which restricts the limiting current of a monoenergetic beam of particles of the same sign of charge in vacuum. Each of these instabilities may, under certain conditions, determine the formation of electric double layers in a plasma.

(6) We devote considerable space in this book to describing *experimental data on limiting currents in beams*. It is shown that, as a result of instabilities, the limiting current of the originally neutralized electron beam is only a few times greater than the limiting current of electrons in vacuum (in the absence of neutralization of their space charge). In the presence of the plasma, the limiting current increases in proportion to its concentration and may exceed the limiting vacuum current by many orders of magnitude.

(7) The method, presented in this book, of analysing the thresholds (critical currents) of the strongest beam instabilities is a very efficient method of testing a theory.

(8) Chapter 8 of this book is devoted to describing the strongest instabilities of ion beams. These instabilities cause violation of the neutrality of the beam space charge. The same chapter describes efficient methods of suppressing these instabilities.

(9) The book also treats certain phenomena in which the plasma *per se* is absent (or may be absent). These phenomena rightly belong to physical electronics; in physical principles, however, they are analogous to certain phenomena in beam plasma; hence, they are conveniently treated from the standpoint of the physics of beams in a plasma. Among such phenomena are, for example, the interactions which lie at the foundation of the newest methods of generation and amplification of electromagnetic waves—by which I mean *'free electron lasers (masers)'*. In this field, the collective effects that are most interesting from the point of view of physics and applications, take place in the regime of hydrodynamic beam instabilities. They are realized in beams with sufficiently high density, satisfying a criterion which is physically equivalent to (1.1) but is quantitatively less severe, namely:

$$\lambda \gtrsim r_{\mathrm{DT}} = (T/4\pi n_1 e^2)^{1/2} \tag{1.3}$$

where λ is the oscillation wavelength and T is the temperature (energy spread) of the beam particles. The principal role in these effects is played by the so-called negative-energy 'waves' which only exist in the hydrodynamic mode of beam instabilities.

(10) Considerable attention is paid in the book to the physical presentation of the concept of negative energy waves. This concept earned a reputation of being quite fruitful not only in plasma physics but also in hydrodynamics, solid state physics, astrophysics and in some other fields. The clarity and usefulness of this concept are demonstrated in the book using examples from the beam physics.

(11) The book also treats the newest experimental data on Langmuir solitons (and other types of soliton that are physically not very different from the Langmuir soliton). The fundamentals of the theory are outlined concurrently, at the 'physical level of rigorousness'. This section of the book is a 'bridge' between the physics of beams in plasma and the most important approaches in modern non-linear plasma physics. The connecting element here is the Langmuir soliton which is formed as a result of the modulational instability (physical collapse) of high-amplitude Langmuir waves generated by the electron beam. These processes are directly related to the mechanism of beam relaxation in a plasma (both in the laboratory and in natural conditions), the mechanisms of particle acceleration in a plasma and the physics of plasma turbulence.

(12) The theory which the first chapters of the book outline is mostly presented in sufficient detail to understand the physics of hydrodynamic beam instabilities.

(13) Chapter 5 of the book is devoted to outlining the methodological and diagnostic aspects of the experimental analysis of beams in plasmas.

The following preliminary remarks are needed to explain the terminology and classification of instabilities in this book. Referring to a *quasineutral beam*, I mean a two-component system consisting of the particles of the beam and an equal amount of neutralizing particles of the opposite sign, without any admixture of 'superfluous' plasma. A three-component system in which this 'superfluous' plasma is present in addition to the beam particles and neutralizing particles, is referred to as a *plasma beam*. Depending on the type of beam particles and the plasma, whose interaction determines a specific instability, I speak about electron–electron instability, electron–ion instability and so forth.

Some sections of this book have been partially presented in my earlier monograph *Beam Dynamics in Plasma* published in Russian by Energoizdat in 1982. Even in these sections, however, this book is considerably more up-to-date and comprehensive: it reflects further experimental and theoretical progress achieved in this field in the last decade, and a substantially updated outlook of the phenomena involved. In comparison with *Beam Dynamics in Plasma*, this book offers completely new sections, such as, for example, double layers in plasma and their relation to the mechanism of polar auroras (chapter 2) and the physical principles that lie at the basis of the modern methods of generating high-power all-wave electromagnetic waves using high-intensity relativistic electron beams (chapter 11). The addition of these new chapters to the book, with a description of currently very important and rapidly developing areas in plasma physics and physical electronics, is a consequence not so much of the author's 'insatiability' as of the pleasant fact that they proved to be closely related to the contents of the preceding chapters of the book and to be expanding considerably the 'sphere of influence' of the physics of hydrodynamical beam instabilities and of its scientific and technical applications.

Presumably, the analysis of instabilities of charged particle beams in plasmas originated with the discovery of the phenomenon identified by the founder of the science of plasma, Irving Langmuir [1.1–1.3]; it is known as the 'Langmuir paradox'. The effect is as follows: a monoenergetic beam of accelerated electrons entering a plasma suddenly undergoes a rapid spreading of the velocity distribution; this happens over distances much shorter than the characteristic length of the corresponding electron–atom and Coulomb collisions; in other words, the relaxation length of an electron beam in the plasma is anomalously short. Furthermore, the electron beam emerging from the plasma contains a significant fraction of particles with energies considerably higher than the energy of electrons entering the plasma. In order to interpret the observational data, Langmuir assumed that these phenomena were caused by the collective interactions of the beam with plasma oscillations (also discovered by Langmuir [1.2]) which the beam itself excites in the plasma. Even paying due respect to Langmuir's powerful intuition, we should remark that the conclusion of the paradoxical nature of the result he obtained had not been immediately accepted. Many researchers were working at that moment with the physics of collisional plasmas, that is, with low-voltage discharge (at beam electron energy from 15 to 30 eV) in relatively high-density, weakly ionized gases; these were conditions in which even the 'normal' relaxation length of the electron beam (explained by the theory of electron–atom pair collisions) was sufficiently short and agreed, within an order of magnitude, with the experimentally observed length. Hence, the Langmuir paradox remained for some time a subject of justified discussions [1.17–1.20]. The situation changed dramatically when physicists moved to analysing the collisionless plasma in rarefied gas pierced by an electron beam of relatively high energy (hundreds or thousands of electron volts). Under these conditions, the paradox became undeniable: the beam relaxation length in the plasma was smaller by several orders of magnitude than that predicted by the theory of pair collisions.

It is interesting to note that plasma researchers, including Langmuir himself, excluded the possibility of explaining the Langmuir paradox by the effect of electron oscillations in electron beams, even a short while before the Langmuir plasma oscillations were discovered (the existence of oscillations in electron beams was known from the physics of microwave electronic devices) [1.1]. The reason for this interpretation was based on the notion reigning at that time that in order to explain the anomalous scattering of electrons in a beam crossing a plasma, it is necessary for the amplitude of the plasma potential oscillations (in energy units) to be very high, of the order of the observed spread of beam electron energies (tens of per cent of the electron beam energy). Since no one observed such intense oscillations, researchers concluded that those they did observe (oscillations of substantially lower amplitude) had nothing to do with the Langmuir paradox. Further arguments in favour of this conclusion are of special interest since they demonstrate, as if embodied in Langmuir's exceptional personality, the process of maturation of plasma physics.... When studying the possible links between the high frequency oscillations and the paradox, Langmuir argued as follows. If such oscillations do happen in the space through which the beam moves, the oscillations must be observable, first, in the oscillations of the discharge current (that is, as voltage oscillations across an inductance specially introduced into the circuit); second, in the HF radiation from the discharge due to current oscillations. However, high-intensity plasma oscillations have not been detected by these methods, so Langmuir concluded that they did not exist [1.1]. Then Langmuir suggested what to him at the time looked a 'more realistic' alternative to explaining the paradox: a hypothetical modification of a collective Compton effect in the system consisting of excited atoms, photons and the electron beam; its probability was to exceed the probability of the ordinary Compton effect by (hold on to your hats!) sixteen orders of magnitude! Only several years later, in a new paper [1.3], Langmuir, having understood the physics of longitudinal plasma oscillations (currently known as Langmuir plasma oscillations) realized that even weak plasma oscillations† (in which the amplitude of the potential is very small in comparison with the loss of energy by beam electrons) can be responsible for a highly efficient energy exchange with the beam [1.3, 1.14]. As for the failure of the detection of oscillations by the methods listed above, it has an elementary explanation in view of new data: the displacement current in longitudinal plasma oscillations is equal, and opposite in sign, to the conduction current; there is nothing surprising, therefore, in that these oscillations cannot be detected—in a first approximation—neither in current oscillations in the external circuit nor in electromagnetic radiation, since they are vortex-free (curl $H = 0$). After Langmuir and Tonks grasped this interpretation of the phenomena [1.3], a qualitative interpretation of the Langmuir paradox met with no major obstacles. Nevertheless, the mechanism of excitation of Langmuir oscillations in the plasma (i.e., of producing a beam instability) was understood only much later (see chapters 3 and 4). Langmuir's

---

† Not the standing oscillations but those propagating as waves whose phase velocity is close to the velocity of the beam electrons.

hypothesis of the decisive role of plasma oscillations in the scattering of the electron beam was conclusively confirmed by recent experimental and theoretical investigations. It became clear afterwards that the Langmuir paradox is caused by the electron–electron instability of the beam in a plasma. This instability is treated in chapters 3 and 4.

Another 'paradox', which at first glance contradicts Langmuir's paradox, has been recently discovered: it was found that if the Langmuir wave amplitude is sufficiently high, then, under certain additional conditions, the beam relaxation length in a plasma increases again (for instance, by several orders of magnitude!); this happens with beams of the so-called 'solar wind', moving at a velocity of about one half of the speed of light [1.21]. The solution of this paradox lies in that the high-amplitude Langmuir waves undergo non-linear self-compression (modulational instability) and group into solitons (see chapter 10). Their wave numbers then increase, the phase velocities decrease, the waves break out of the Čerenkov resonance with the beam, and the plasma becomes much more 'transparent' for the beam. This phenomenon may occur when the increment of the modulational instability is greater than the increment of the beam instability [1.22].

In connection with the physics of Langmuir waves, it is interesting to mention the following unusual object: the so-called stationary recombining 'supercooled' plasma produced by a high-intensity electron beam in a high-density gas [1.23, 1.24]. The state of this unusual medium differs from thermodynamic equilibrium in exceptionally high concentrations of electrons and excited atoms, at a relatively low electron temperature: $T_e = 0.1$–0.2 eV. For this medium to exist (which is of interest for the physical and chemical kinetics of excited atoms and molecules), it is necessary to eliminate any possibility of triggering a beam instability capable of heating plasma electrons.

After these preliminary remarks, we can start an analysis of specific beam instabilities.

References

[1.1]   Langmuir I 1925 *Phys. Rev.* **26** 585
[1.2]   Langmuir I 1928 *Proc. Nat. Acad. Sci. (Washington)* **14** 627
[1.3]   Tonks L and Langmuir I 1929 *Phys. Rev.* **33** 195
[1.4]   Pierce J R 1948 *J. Appl. Phys.* **19** 231
[1.5]   Pierce J R 1944 *J. Appl. Phys.* **15** 721
[1.6]   Pierce J R 1950 *Travelling Wave Tubes* (New York: Van Nostrand)
[1.7]   Pierce J R 1947 *Proc. IRE* **35** 111
[1.8]   Haeff A V 1948 *Phys. Rev.* **74** 1532
[1.9]   Haeff A V 1949 *Proc. IRE* **37** 4

[1.10] Akhiezer A I and Fainberg Ya B 1949 *Doklady AN SSSR* **69** 555

[1.11] Akhiezer A I and Fainberg Ya B 1951 *Zh. Eksp. Theor. Fiz.* **21** 1262

[1.12] Fainberg Ya B 1961 *Atomnaya Energiya* **11** 313 (Engl. Transl. 1962 *Sov. J. At. Energy* **11** 958)

[1.13] Fainberg Ya B 1978 *Ukr. Fiz. Zh.* **23** 1885

[1.14] Bohm D and Gross E P 1949 *Phys. Rev.* **75** 1851, 1864

[1.15] Kadomtsev B B 1980 *Collective Phenomena in Plasmas* (Oxford: Pergamon Press); (1989) 2nd edition (Moscow: Nauka)

[1.16] Artsimovich L A and Sagdeev R Z 1979 *Plasma Physics for Physicists* (Moscow: Atomizdat) (in Russian)

[1.17] Nezlin M V 1982 *Beam Dynamics in Plasma* (Moscow: Energoizdat) (in Russian)

[1.18] Granovsky V L 1952 *Electric Current in the Gas* (Moscow: Atomizdat) ch 8, sec 53 (in Russian)

[1.19] Crawford F W and Self S A 1965 *Int. J. Electronics* **18** 569

[1.20] Sena L A 1948 *Collisions of Electrons and Ions with Gas Atoms* (Moscow–Leningrad: Gostekhteorizdat) (in Russian)

[1.21] Goldman M V 1984 *Rev. Mod. Phys.* **56** 709

[1.22] Galeev A A, Sagdeev R Z, Shapiro V D and Shevchenko V I 1977 *Zh. Eksp. Theor. Fiz.* **72** 507 (Engl. Transl. 1977 *Sov. Phys.-JETP* **45** 266)

[1.23] Antipov S V, Nezlin M V, Snezhkin E N and Trubnikov A S 1973 *Zh. Eksp. Theor. Fiz.* **65** 1866 (Engl. Transl. 1974 *Sov. Phys.-JETP* **38** 931)

[1.24] Snezhkin E N and Nezlin M V 1977 *Zh. Eksp. Theor. Fiz.* **73** 913 (Engl. Transl. 1977 *Sov. Phys.-JETP* **46** 481)

2   Aperiodic Beam Instabilities (Theory)

This chapter treats instabilities which transform a beam to a new state; they are of a loss-of-equilibrium type. First and foremost among them are the Bursian and the Pierce instabilities.

2.1   Bursian Instability and Limiting Current of a Beam of Particles of Identical Sign in Vacuum

Let a monoenergetic beam of particles of identical sign (assume it to be an electron beam) propagate in an equipotential space bounded by metal walls. Experimentally the beam can be obtained by, for example, a 'gun' of some sort, and be sent along the axis of a metal cylinder whose end faces are covered by transparent grids (figure 2.1). For the sake of simplification, we assume that the beam is shaped into a homogeneous rod whose diameter, $2a$, is much less than the diameter $2R$ of the metal cylinder enclosing the beam. This geometry can actually be implemented by placing the beam in a strong longitudinal magnetic field. In what follows, we always assume (unless the opposite is specified) that the electron beam propagates in a strong longitudinal magnetic field H in which the Larmor frequency is higher than the Langmuir frequency:

$$\omega_H = eH/mc > \omega_1 \qquad (2.1)$$

where

$$\omega_1 = (4\pi n_1 e^2/m)^{1/2} \qquad (2.2)$$

and n_1 is the beam electron density (formulas (2.1), (2.2) are written in the non-relativistic approximation).

If $a \ll R$, we can assume in the case under consideration, that the entire potential drop due to the space charge (in cross sections of the system at sufficient distances from the beam ends) is concentrated outside the beam,

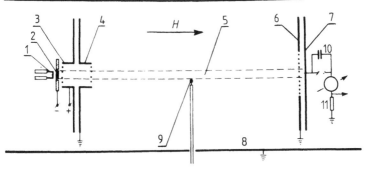

Figure 2.1   Electron beam in equipotential space: 1–4, electron gun: 1, filament; 2, indirectly heated cathode (tungsten); 3, intermediate electrode (grid); 4 and 6, grounded grids; 5, beam; 7, beam collector; 8, metal cylinder; 9, electrostatic probe; 10, antidinatron battery (50 V); 11, measurement resistor. Beam diameter $2a$, cylinder diameter $2R$.

while inside the beam all electrons have identical energies. If the beam current is I and the electron velocity (perturbed by the space charge) is u, the radial electric field at a distance r from the beam axis is $E = 2I/(ru)$ and the potential difference between the surrounding walls and the beam (that is, between the end grids and the midpoint of the beam) is

$$\varphi = \frac{2I}{u} \ln \frac{R}{a}. \tag{2.3}$$

This potential difference decelerates the electrons of the beam so that their velocity decreases:

$$u = \left[\frac{2e}{m}(V_0 - \varphi) \right]^{1/2} \tag{2.4}$$

where $eV_0 = mu_0^2/2$ is the energy of electrons immediately before entry into the system and u_0 is their non-perturbed velocity.

These relations give

$$I = \frac{(2e/m)^{1/2}}{2\ln(R/a)} \, \varphi(V_0 - \varphi)^{1/2}. \tag{2.5}$$

If we also take into account the potential drop inside the beam (assuming the charge density in it to be homogeneous), then instead of (2.5) we obtain

$$I = \frac{(2e/m)^{1/2}}{1 + 2\ln R/a} \, \varphi(V_0 - \varphi)^{1/2}. \tag{2.6}$$

This expression shows that the beam current reaches a maximum at

$$\varphi = \varphi_m = \frac{2}{3}V_0. \tag{2.7}$$

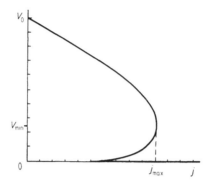

Figure 2.2   Potential on the axis as a function of electron current (the case of planar geometry).

Therefore, the maximum (limiting) current is

$$I_{\max} = \frac{2(2e/m)^{1/2}V_0^{3/2}}{3\sqrt{3}(1 + 2\ln(R/a))}. \tag{2.8}$$

Hence, if we increase the beam current I at the input to the system above the value $I_{\max}$ (this means that $\varphi > \frac{2}{3}V_0$), then the beam current emerging from the system falls below $I_{\max}$. This means that some electrons of the beam do not travel across the system and are reflected back towards the beam source. In other words, whatever the increase in the current above the value $I_{\max}$, the beam forms a virtual cathode, that is, the retarding (beam-decelerating) potential difference increases jumpwise from $\varphi = \frac{2}{3}V_0$ to $\varphi = V_0$ (figure 2.2). This jump is the sign of instability. It was given the name of Bursian, who discovered it experimentally and developed its theory (see [2.1, 2.2]).

The Bursian instability is clearly illustrated with the curve (figure 2.2) which expresses the relation between the potential V at the axis of a cylindrical beam and the beam current I [2.5, 2.6, 2.8]. If the beam current is low, its potential far from the end grids is quite close to the grid potential: $V \simeq V_0 = W_1/e$ (W_1 is the beam electron energy at the entrance to the system). As the current increases, the space charge of the beam also increases (there are no neutralizing ions), so that the beam potential decreases. As long as the beam current is such that $V > V_{\min} = \frac{1}{3}V_0$, the potential V decreases continuously as I increases. However, when $V = V_{\min}$, an increase in I produces a jumpwise drop of V to zero. At this moment, a virtual cathode arises in the beam and the beam current undergoes disruption. In other words, there is an instability at $V = V_{\min}$ with respect to the formation of a virtual cathode. The characteristic time of evolution of instability is of the order of the time of flight of a beam electron across the system, provided $I \gg I_{\max}$.

It is necessary to note here that one part of the beam can pass through the virtual cathode while another part of the beam is reflected from it only if there is a certain velocity spread in the beam. This spread can, however, be arbitrarily small, so that the initial assumption of monoenergetic beam does not violate the correctness of the solution obtained (for details, see [2.3, 2.4]).

If we take into account the potential drop in the beam (we again assume it to be small), it is easy to show that (2.5) is replaced with an expression for the instability threshold (the Bursian limiting current):

$$I_{max} \equiv I_B = \frac{2}{3\sqrt{3}} (2e/m)^{1/2} \frac{V_0^{3/2}}{1 + 2\ln(R/a)} \qquad (2.9)$$

or

$$I_B = 25.4 \times 10^{-6} \frac{V_0^{3/2}}{1 + 2\ln(R/a)} \qquad (2.10)$$

where V_0 is assumed to be in volts and I_B, in amperes. These relations hold if $2\ln R/a \gg 1$. If $R \simeq a$, then [2.5, 2.6]

$$I_B \simeq 32 \times 10^{-6} V_0^{3/2}. \qquad (2.11)$$

We can recast this instability threshold in terms of the Langmuir beam frequency:

$$\omega_{1B} = \omega_1 \quad \text{for} \quad I = I_B \qquad (2.12)$$

$$\omega_{1B}^2 \equiv \frac{4e}{ma^2} \frac{I_B}{u_0} = \frac{1}{3\sqrt{3}} k_\perp^2 u_0^2 \qquad (2.13)$$

where, for $R \gg a$,

$$k_\perp^2 \simeq \frac{2}{a^2 \ln(R/a)} \qquad (2.14)$$

is the characteristic wave number squared.

In an electron beam with non-compensated space charge, a hysteresis may be observed. For instance, if the input current of the system is increased beyond the Bursian limit, the beam current emerging from the system decreases: the beam is shut off by the virtual cathode. If one now decreases the input current of the system, the beam opens up not at $I = I_B$ but at a considerably lower current, namely at $I = I_B/2$ (for simplicity, we consider the planar geometry). Therefore, if the current is controlled as described above, the state of the system follows a hysteresis loop shown in figure 2.3. The virtual cathode then displays spatial oscillations: as the current increases to $I = I_B$, the virtual cathode first forms at the midpoint

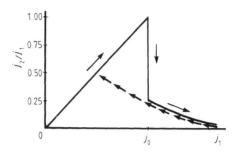

Figure 2.3   Beam current output of the system as a function
of the input current (for planar geometry).

of the interelectrode space, and shifts towards the beam source if the input
current $I > I_B$.

Intuitively, it seems probable that the virtual cathode may also display
spatial vibrations in the case when the input beam current is maintained
constant. Indeed, this prediction was confirmed: oscillations of the virtual
cathode in electron beams with over-Bursian current do exist and are used
in very high-power modern microwave generators, the so-called vircators
discussed in chapter 11.

In the theoretical paper [2.7], the current disruption of a non-
compensated electron beam due to the Bursian instability is deduced from
the soliton dynamics of space charge; these solitons are discussed in section
10.6.

2.2   Instability and Limiting Current of Quasineutral (Compensated) Electron Beams

2.2.1   Pierce problem: compensating ions are at rest

Pierce proved theoretically in 1944 [1.5, 2.8] that the beam current cannot
exceed a certain limit—only several times the Bursian current (2.9),
(2.11)—even if the space charge of beam electrons is neutralized by positive
ions (which were assumed to be at rest). This phenomenon is also caused
by an instability. Its mechanism is the same as in the Bursian instability;
Pierce demonstrated that it is related to a positive feedback realized via
electrons of the external circuit which sustain the equipotentiality of the
grids and walls bounding the beam. If the beam current exceeds a certain
limit (the Pierce threshold I_P), then the condition of equipotentiality of the
grids coincides with the condition of instability of the system with respect
to increasing depression of the potential in the space between the grids.
This condition of instability can be obtained from the following simple
arguments.

Let a negative fluctuation of the potential, $\delta\varphi_1$, occur in the space between the grids (in a cylindrical geometry like that shown in figure 2.1). It decreases the velocity of beam electrons by an amount

$$\delta u = (e/2m)^{1/2} V_0^{-1/2} \delta\varphi_1. \tag{2.15}$$

As a result, the space charge I/u per unit length of the beam increases, at a given beam current, by the amount

$$\delta q = \frac{I}{u^2}\delta u \tag{2.16}$$

which causes further reduction of the spatial potential by a quantity

$$\delta\varphi_2 = (1 + 2\ln(R/a))\delta q = \frac{1}{2}(1 + 2\ln(R/a))\left(\frac{2e}{m}\right)^{-1/2} V_0^{-3/2}\delta\varphi_1.$$

The process of decrease of the spatial potential is steadily enhanced if $\delta\varphi_2 > \delta\varphi_1$. A non-linear analysis then shows that the process continues until a virtual cathode is formed [2.9–2.11]. Now we can find the threshold current of the aperiodic Pierce instability that we are discussing (the result is confirmed by a more rigourous calculation [2.8, 2.9, 2.12, 2.13]):

$$I_P = 2(2e/m)^{1/2}\frac{V_0^{3/2}}{1 + 2\ln(R/a)} = 3\sqrt{3}\,I_B \tag{2.17}$$

$$I_P \simeq 133 \times 10^{-6}\frac{V_0^{3/2}}{1 + 2\ln(R/a)} \tag{2.18}$$

(here the voltage V_0 is given in volts and the current I is in amperes) and

$$\omega_{1P}^2 \equiv \frac{4e}{ma^2u_0}I_P = k_\perp^2 u_0^2. \tag{2.19}$$

The derivation above shows the difference between the Pierce and the Bursian thresholds: in the non-relativistic approximation that we are discussing, the former is greater than the latter by a factor of $V_0^{3/2}/(V_0 - \varphi_m)^{3/2} = 3\sqrt{3}$.

The relations derived (they are valid for $2\ln R/a \gg 1$) show that if the beam current exceeds the Pierce threshold, the criterion of a high-intensity beam ((1.1) and (1.2)) is satisfied; thus, the characteristic transverse beam size $l_\perp \simeq 1/k_\perp$ exceeds its Debye radius.

In the case of one-dimensional (planar) geometry, the Pierce instability condition takes the form

$$\left.\begin{array}{c} \omega_1 \geqslant \omega_{1P} = k_z u_0 \\ \text{or} \qquad \alpha \equiv \omega_1 L/u_0 \geqslant \pi \end{array}\right\} \tag{2.20}$$

where $k_z = \pi/L$ and L is the beam length.

The Pierce instability increment γ, that is, the characteristic inverse time of exponential growth of a small negative fluctuation of spatial potential, is, in the case of sufficiently high overcriticality (e.g., if $I \simeq 2I_p$), of the order of the inverse time of flight, u_0/L. Since we always have $\gamma \propto 1/L$, this instability is characteristic (as the Bursian instability is) only of finite length beams: $\gamma \to 0$ as $L \to \infty$.

To conclude this section, several remarks are in order. First, we have assumed so far that ions compensating the space charge of beam electrons are at rest (i.e., their mass $M \to \infty$). It was shown [2.14, 2.15] that if the ion mobilities are taken into account ($M \neq \infty$), the instability does not vanish although it acquires certain specific features (for details, see sections 3.1 and 3.6). What I described above was the evolution of a *negative* fluctuation of the potential of a quasineutral electron beam. If the fluctuation is positive, the system again displays an instability: as a result of acceleration of beam electrons by the positive potential fluctuation, their space charge decreases which further increases the spatial potential, and so on. This instability results in strong positive surges of potential, which accelerate compensating ions to substantial energies [2.16–2.18] (for details, see subsection 2.2.3). Second, so far we have been discussing a two-component quasineutral beam, composed of fast electrons and charge-compensating ions (the system has no slow electrons). If additional plasma is introduced into such a beam, that is, if the system is made three-component, then the limiting beam current increases substantially, owing to an efficient screening of the electric field of the fluctuation by plasma electrons. This effect is discussed in section 6.5.

2.2.2   Stabilization of Pierce and Bursian instabilities by velocity spread in the beam

The basic assumption of the preceding sections was that the electron beam was monoenergetic. The situation may change qualitatively if the beam has a considerable velocity spread. In order to show this, let us consider a beam with a stepwise distribution function of electron energies (figure 2.4); as before, the beam is neutralized by ions at rest. In this case, if a negative fluctuation of potential $\delta\varphi$ arises, the space charge of the beam electrons must, on one hand, increase because of deceleration, and on the other hand, it must decrease because the beam current decreases. Namely, by virtue of (2.16) and figure 2.4,

$$\delta q = \frac{I}{u_{av}^2}\delta u - \delta I/u_{av}$$

where, for the distribution function chosen (figure 2.4), $\delta I = I\delta\varphi/V_0$ and $u_{av} = (eV_0/2m)^{1/2}$ is the average velocity of beam electrons.

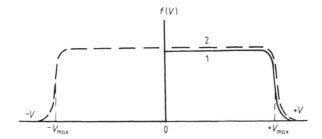

Figure 2.4   Electron beam with 'stepwise' electron velocity distribution function.

Expressing δu in terms of $\delta \varphi$ via (2.15), we find that in our example

$$\delta q = 0$$

so the system is stable.

Therefore, a beam with a sufficiently large spread of electron velocities does not display the Pierce instability. The same can be said, according to section 2.1, about the Bursian instability because its physical mechanism is the same.

A different approach to analysing the effect of velocity spread on the beam instability was applied in the theoretical papers [2.19–2.21]. These papers showed that the Pierce instability exists (and has positive increment $\gamma > 0$) only for the 'Mach number' $M > 1$, where $M = u_0/\Delta u$ and Δu is the characteristic velocity spread of the beam (we assume that the beam electron distribution function has the form $u = u_0 \exp(-(u_0 - u)^2/(\Delta u)^2))$. According to [2.19–2.21], if $\Delta u \simeq u_0$, then $\gamma \to 0$.

This result [2.21] is illustrated in figure 2.5 which plots the Pierce instability increment (in units of ω_1) as a function of a parameter p, characterizing both the instability threshold and the Mach number: for planar geometry

$$p = \alpha\beta^{1/2} \quad \text{where} \quad \alpha = \frac{\omega_1 L}{u_0} \quad \text{and} \quad \beta = \frac{M^2}{M^2 - 1}.$$

We see that as the velocity spread of the beam increases ($M \to 1$), the instability increment $\gamma \to 0$, the instability vanishes.

2.2.3   Soft mode of Pierce instability: beam current not disrupted, ions accelerated

The Pierce instability regime described in the preceding subsection corresponds to the following situation: if the beam current I in the initial state of the system is greater than the instability threshold I_P, then the new state of the system (the one with the virtual cathode) is independent of

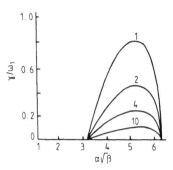

Figure 2.5 Effect of velocity spread of beam electrons (Δu) on the aperiodic Pierce instability increment in the case of ions at rest. Here $\beta = M^2/(M^2 - 1)$, where $M = u_0/\Delta u$ is the 'Mach number'. Numbers at the curves give the values of β.

the degree of overcriticality, $(I - I_P)$. This is known as the rigid instability regime. A system reaches this regime with an over-Pierce electron beam current in response to a negative fluctuation of the spatial potential.

In contrast to this rigid regime, an instability regime may be soft. Namely, the final state of the system is then determined by the degree of its overcriticality in the initial state. The system reaches this regime in response to a positive potential fluctuation.

The difference between these two regimes is dictated by non-linear processes. Indeed, in the linear approximation, the fluctuation of the spatial potential of an unstable system causes a corresponding change in the velocity of beam electrons which, for a given beam current, changes their density (their space charge); as a result, the deviation of the spatial potential from the initial value increases, regardless of the sign of this deviation.

In a non-linear regime, the system's evolution may be drastically different in response to a negative or positive potential fluctuation. This is fairly easy to show if we consider a beam whose current is only slightly greater than the Pierce threshold. If a negative fluctuation occurs in this beam, it grows faster and faster as a result of a progressive (non-linear) decrease in the velocity of electrons, that is, as a result of the growth in their space charge. Consequently, the system evolves to a state with a virtual cathode [1.5, 2.9, 2.11]. If a positive potential fluctuation occurs, then the Pierce threshold increases in proportion to u^3 in response to the non-linear increase in the electron velocity. Hence, if $I = $ constant, a moment comes when the beam current ceases to exceed the Pierce threshold, so that the instability vanishes. As a result, the potential of the system increases in proportion to the degree of overcriticality of the system, $\Delta I = I - I_P$. This is known as the soft instability regime. If $\Delta I \gg I_P$, the system's potential in the final state may rise to a level much greater than the beam electron energy W_1 in the initial state:

$$e\varphi \simeq \frac{\Delta I}{I_{\rm P}} W_1 \gg W_1. \tag{2.21}$$

A more detailed analysis shows [2.16–2.18] that if $\Delta I \gg I_{\rm P}$, a beam state is possible with oscillating potential distribution along the system: it has numerous potential maxima and virtual cathodes.

In reality, this state can exist for a limited time only, namely, not longer than the time of recession of the compensating ions in the field of the positive spatial potential [2.17].

2.2.4 Positively charged electron beams with raised Pierce instability threshold

Imagine a system in which the longitudinal distribution of potential differs from that shown in figure 2.1. Namely (figure 2.6), the electron energy equals $W_{10} = eV_0$ immediately before entry into the drift space (i.e. slightly to the right of the input grid in the anode) and on emergence from it (i.e. slightly to the right of the output grid in the auxiliary cylinder), but is considerably higher at the midpoint of the drift space. This situation arises if there is an excess space charge of positive ions within the drift space, above the charge needed to neutralize the space charge of the beam electrons [2.22, 2.23]. In this case, the Pierce instability does develop but its threshold is now determined by the energy that electrons acquire at the midpoint of the interelectrode space ($W_1 = W_{10} + \Delta W$). The latter energy is greater than W_{10} by a quantity which corresponds to the acceleration of beam electrons by the excess charge of positive ions. The instability threshold corresponds again to formula (2.17) in which the quantity W_1 plays the role of eV_0. This threshold is greater by a factor of $(W_1/W_{10})^{3/2}$ than the current which is determined by (2.17) at $eV_0 = W_{10}$. In this sense, we can say that the system allows a stable current greater than the Pierce threshold which is determined by the initial electron energy W_{10}. However, the stable limiting beam current does not exceed the Pierce threshold which is determined by the electron energy W_1. In this last sense, stable beams carrying 'over-Pierce' current are forbidden.

With these qualifications, it can be said that positively charged electron beams promise interesting applications [2.22]. One of the simpler methods of positively charging a beam is to create a three-component system (by adding an 'excessive' plasma to a quasineutral beam) and purge this system of slow electrons (i.e., plasma electrons). This purging can be done by surrounding the beam on all sides by two grids inserted one into another, and connecting the outer grid (or electrode) to a potential that is positive with respect to the inner grid. If the beam ionizes the gas in the drift space, it can serve as its own plasma source [2.23] (see figure 2.6).

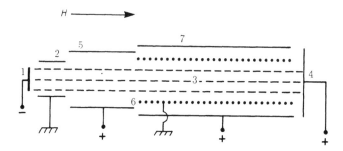

Figure 2.6 Potential distribution along the axis of a positively charged electron beam. 1, cathode (beam source); 2, anode (transparent for electrons); 3, beam; 4, beam collector; 5, auxiliary cylinder which determines high positive potential of the beam; 6, grid; 7, positive electrode.

Let us consider in more detail the relation between the limiting electron beam current and the space charge of positive ions. We denote the ion charge per unit length of the beam by

$$Q_+ = \pi a^2 e n_+$$

where n_+ is the ion concentration. We assume that this charge is connected with the beam electron charge per unit length, $Q_- = I/(2eV/m)^{1/2}$, via an arbitrary degree of beam compensation $f \geqslant 0$, namely: $Q_+ = f \cdot Q_-$. The situation corresponds to the Bursian problem if $f = 0$, and to the Pierce problem if $f = 1$. Let us generalize these problems to the case of arbitrary f. Note, first of all, that the charge Q_+ raises the midpoint potential of the beam relative to the walls by a quantity

$$\Delta V = Q_+(1 + 2\ln(R/a)). \tag{2.22}$$

Therefore, instead of (2.6) the relation between the beam current and the spatial potential becomes

$$I = \frac{(2e/m)^{1/2}V^{1/2}}{(1 + 2\ln(R/a))}\left[V_0 + Q(1 + 2\ln(R/a)) - V\right]. \tag{2.23}$$

In the Bursian problem, the maximum of the beam current was reached at $V = \frac{1}{3}V_0$; similarly, in the present case the beam current reaches maximum at $V = \frac{1}{3}[V_0 + Q_+(1 + 2\ln R/a)]$. Now the maximum beam current is

$$
\begin{aligned}
I_{max}(Q_+) &= \frac{2(2e/m)^{1/2}}{3\sqrt{3}}\frac{[V_0 + Q_+(1 + 2\ln(R/a))]^{3/2}}{1 + 2\ln(R/a)} \\
&= I_0\left[1 + \frac{Q_+(1 + 2\ln(R/a))}{V_0}\right]^{3/2}
\end{aligned}
\tag{2.24}
$$

where $I_0 = I_B$ is the Bursian limiting current. Thus, if $Q_+(1 + 2\ln R/a) = 2V_0$ then $I_{max} = 3\sqrt{3}I_0 = I_P$; if the quantity Q_+ is still greater, the limiting beam current becomes 'over-Pierce' in the sense outlined above.

Equations (2.23) and (2.9) show that at the limiting current we have

$$Q_- = \frac{2}{3}\frac{V_0 + Q_+(1 + 2\ln(R/a))}{1 + 2\ln(R/a)}.$$

Taking into account the identity

$$Q_+ = Q_+ \frac{1 + 2\ln(R/a)}{1 + 2\ln(R/a)}$$

we find that at the limiting current, the degree of compensation of the space charge of beam electrons by the charge of positive ions is

$$f_{max} = \frac{Q_+}{Q_-} = \frac{3}{2}\frac{Q_+(1 + 2\ln(R/a))}{V_0 + Q_+(1 + 2\ln(R/a))}. \tag{2.25}$$

We conclude that when Q_+ varies from 0 to ∞, the quantity f changes from 0 to 3/2: when the beam current equals the limiting current, the degree of compensation due to ions does not exceed 3/2.

The relations given here imply the following relation of the limiting current to the Bursian current:

$$I_{max} = \frac{I_B}{(1 - \frac{2}{3}f_{max})^{3/2}}.$$

We find that $I_{max} = I_B$ for $f_{max} = 0$, and $I_{max} = 3\sqrt{3}I_B = I_P$ for $f_{max} = 1$. If $f_{max} \rightarrow \frac{3}{2}$, then $I_{max} \rightarrow \infty$. This last result is not surprising since (2.25) shows that in order to obtain $f_{max} \rightarrow \frac{3}{2}$, we need $Q_+ \rightarrow \infty$, that is, an infinitely high beam current requires an infinitely high charge of positive ions; this also implies infinitely high energy W_1 acquired by the beam electrons in the field of the space charge of positive ions.

These conclusions are illustrated in figure 2.7 which generalizes the result of figure 2.2 to the case of $Q_+ > 0$. In figure 2.7, curve 1 corresponds to the Bursian regime, curve 3 to the Pierce regime, and curve 4, to the 'over-Pierce' limiting current.

As for the experimental realization of 'over-Pierce' currents, the limiting beam current that has been reported so far [2.23] exceeds the 'Pierce' current dictated by $V_0^{3/2}$ by a factor of at most 1.5. This modest result is determined by the difficulty of creating a sufficiently high charge Q_+, owing to leaking of ions to the nearest of the two grids surrounding the beam [2.23].

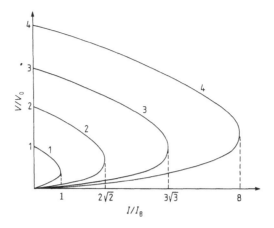

Figure 2.7   The potential on the cylinder axis as a function of beam current, for various values of the charge of positive ions per unit length, Q_+. 1, $Q_+ = 0$, $I_{max} = I_B$ (Bursian regime); 2, $Q_+ = Q_1 = V_0/2\ln R/a$, $I_{max} = 2\sqrt{2}I_B$; 3, $Q_+ = 2Q_1$, $I_{max} = 3\sqrt{3}I_B = I_P$; 4, $Q_+ = 3Q_1$, $I_{max} = 4\sqrt{4}I_B = 8I_B$.

2.3   Instability and Limiting Current of Quasineutral Ion Beams

Let us consider the problem of the stability of an intense ion beam whose space charge is electron-compensated. At first glance, this problem seems to be completely analogous to the Pierce problem. Actually, however, the analogy is rather superficial because the charge of the heavy ions of the beam is compensated in this case by light particles (electrons) which are highly mobile, in contrast to fixed ions in the Pierce problem.

Assume now that compensating electrons are inertialess. It is not difficult to see that this factor introduces a new dimension to the problem. Now, if a primary potential fluctuation, which decelerates the beam, occurs in this beam, the high-mobility electrons rapidly fill the potential well formed in the beam. These electrons place the entire non-compensated charge of the beam (which results from the deceleration of ions) in those layers of potential drop which always exist at the boundaries of a steady-state beam. The potential drop in the boundary layer is dictated by the condition of confining the necessary amount of compensating electrons in the beam; its value equals several times the electron temperature (T_e/e) and the layer thickness equals several electron Debye radii. It is clear that the non-compensated charge transported to the thin boundary layer creates a much smaller potential difference—which decelerates ions—than it creates in the case when this charge is almost uniformly distributed along the entire length (width, height) of the beam. Therefore, the instability threshold increases appreciably and, as can be shown [2.24–2.26], is given by the expression

$$I_c \simeq \frac{W_1}{T_e} \left(\frac{2e}{M}\right)^{1/2} \frac{V_0^{3/2}}{L^2} \qquad (2.26)$$

where W_1 is the ion energy, M is the ion mass and L is the beam length (we consider the one-dimensional case). If the beam current is greater than the critical current I_c, the instability must lead to the formation of a virtual anode in the beam; this virtual anode reflects the primary ions of the beam. In other words, the current I_c must be the limiting current of the beam. In comparison with the similar Pierce instability, and taking into account the 'ionic' ratio e/M, this instability has a considerably higher threshold (approximately by a factor of W_1/T_e).

2.4 Instability and Limiting Current of Relativistic Electron Beams

If a beam is strongly relativistic, then $u \simeq c$ regardless of the potential depression due to space charge, so that the denominator of the limiting Bursian current does not contain the factor $3\sqrt{3}$:

$$I_B \simeq \frac{ma^2}{4e} k_\perp^2 c^3 \gamma_0 \simeq \frac{mc^3}{e} \frac{\gamma_0}{1 + 2\ln(R/a)} \qquad (2.27)$$

or

$$I_B \simeq 17\gamma_0 \frac{1}{1 + 2\ln(R/a)} \, kA \qquad (2.28)$$

where

$$\gamma_0 = \frac{\mathcal{E}}{mc^2} = 1 + \frac{eV_0}{mc^2} \qquad (2.29)$$

is the relativistic factor, $\mathcal{E} = mc^2 + eV_0$ is the total electron energy and mc^2 is the rest energy of an electron.

A simple interpolation formula holds for arbitrary energy of beam electrons [2.27]:

$$I_B = \frac{mc^3}{e} \frac{(\gamma_0^{2/3} - 1)^{3/2}}{1 + 2\ln(R/a)}. \qquad (2.30)$$

If the electron velocity is low or high, this formula transforms into respective relations which correspond to the limiting cases discussed above.

The Pierce instability threshold increases by a factor of γ_0^3 (see [2.8]):

$$I_P = \frac{mc^3}{e} \frac{\gamma_0^3}{1 + 2\ln(R/a)}. \qquad (2.31)$$

Note that relations (2.27–2.31) hold only if condition (2.1) of the magnetization of beam electrons by the external longitudinal magnetic field is satisfied. If this condition is violated, decreasing magnetic field raises the Pierce instability threshold, as a result of a strong divergence of electron trajectories caused by the repulsion among electrons [2.28, 2.29]. In the case of relativistic beams, the condition of the magnetization of electrons is written as

$$\frac{H_0^2}{8\pi} > \gamma_0 n_1 mc^2. \tag{2.32}$$

Finally, it is necessary to remark that in the case of the relativistic-beam regime, referring to the Pierce instability with ions at rest is justified only in a very limited sense: electrons in longitudinal oscillations get 'heavier' by a factor of γ_0^3, so that the assumption of ions at rest at $\gamma_0 \simeq (M/m)^{1/3}$ (and even at lower beam energies) ceases to be correct and we also need to take the ion mobility into account.

References

[2.1]   Bursian V R 1921 *Zh. Russk. Fiz.-Khim. Obsh* **51** 289

[2.2]   Bursian V R and Pavlov V I 1923 *Zh. Russk. Fiz.-Khim. Obsh.* **55** 71

[2.3]   Dobretsov L N 1950 *Electron and Ion Emission* (Moscow: Gostekhizdat) Sect. 28 (in Russian)

[2.4]   Bellyustin S V 1939 *Zh. Eksp. Theor. Fiz.* **9** 742

[2.5]   Smith L P and Hartman P L 1940 *J. Appl. Phys.* **11** 220

[2.6]   Pierce J R 1954 *Theory and Design of Electron Beams* (New York: D. Van Nostrand) Chapter 9

[2.7]   Rutkevich B N and Rutkevich S B 1990 *Fizika Plazmy* **16** 683 (Engl. Transl.1991 *Sov. J. Plasma Phys.* **16** 608)

[2.8]   Nezlin M V 1970 *Uspekhi Fiz. Nauk* **102** 105 (Engl. Transl. 1971 *Sov. Phys.-USPEKHI* **13** 608)

[2.9]   Smirnov V M 1966 *Zh. Eksp. Theor. Fiz.* **50** 1005 (Engl. Transl. 1966 *Sov. Phys.-JETP* **23** 668)

[2.10] Kuzelev M and Rukhadze A 1990 *Electrodynamics of Dense Electron Beams in Plasma* (Moscow: Nauka)

[2.11] Shapiro V D and Shevchenko V I 1967 *Zh. Eksp. Theor. Fiz.* **52** 144 (Engl. Transl. 1967 *Sov. Phys.-JETP* **25** 92)

[2.12] Frey J and Birdsall C K 1966 *J. Appl. Phys.* **37** 2051

[2.13] Karbushev N I and Rukhadze A A 1987 in *Generators and Amplifiers on Relativistic Electron Beams* (Moscow: Moscow Univ. Publ) p 154 (see also references therein)

[2.14]   Vladimirov V V, Mosiyuk A N and Mukhtarov M A 1983 *Fizika Plazmy* **9** 992 (Engl. Transl. 1983 *Sov. J. Plasma Phys.* **9** 578)

[2.15]   Kolyshkin I N, Kuznetsov V I and Ender A Ya 1984 *Zh. Tekhn. Fiz.* **54** 1512 (Engl. Transl. 1984 *Sov. Phys. Tech. Phys.* **29** 882)

[2.16]   Burinskaya T M and Volokitin A S 1983 *Fizika Plazmy* **9** 453 (Engl. Transl. 1983 *Sov. J. Plasma Phys.* **9** 261)

[2.17]   Burinskaya T M and Volokitin A S 1984 *Fizika Plazmy* **10** 989 (Engl. Transl. 1984 *Sov. J. Plasma Phys.* **109** 567)

[2.18]   Crystal T L and Kuhn S 1985 *Phys. Fluids* **28** 2116

[2.19]   Yan K 1977 *Phys. Fluids* **48** 133

[2.20]   Mosiyuk A N 1986 *Fizika Plazmy* **12** 1493 (Engl. Transl. 1986 *Sov. J. Plasma Phys.* **11** 864)

[2.21]   Kuhn S 1987 in *Intern. Conf. on Plasma Physics* (Kiev-87) Invited papers vol **2** (Singapore: World Scientific) p 954

[2.22]   Vybornov S I, Zharinov A V and Malafaev V A 1988 *Fizika Plazmy* **14** 84 (Engl. Transl. 1988 *Sov. J. Plasma Phys.* **14** 51)

[2.23]   Malafaev V A 1990 *Fizika Plazmy* **16** 1085 (Engl. Transl. 1990 *Sov. J. Plasma Phys.* **16** 629)

[2.24]   Popov Yu S 1966 *Pis'ma v Zh. Eksp. Teor. Fiz.* **4** 352 (Engl. Transl. 1966 *JETP-Lett.* **4** 238)

[2.25]   Zharinov A V 1973 *Pis'ma v Zh. Eksp. Teor. Fiz.* **17** 508 (Engl. Transl. 1973 *JETP-Lett.* **17** 366)

[2.26]   Erofeev V S and Leskov L V 1974 in *Physics and Application of Plasma Accelerators* (Minsk: Nauka i Tekhnika) p 23 (in Russian)

[2.27]   Bogdankevich L S and Rukhadze A A 1971 *Uspekhi Fiz. Nauk* **103** 609

[2.28]   Ivanov A A and Putvinskaya N S 1975 *Zh. Tekhn. Fiz.* **45** 1648

[2.29]   Ignatov A M and Rukhadze A A 1984 *Fizika Plazmy* **10** 112

3 Oscillatory Beam Instabilities (Theory)

3.1 Oscillatory Pierce Instability and its Evolution to Chaos: from Regular Oscillations via a Sequence of Period-Doubling Bifurcations to a Stochastic Attractor and its Crisis

In the preceding chapter, we discussed the Pierce instability close to the minimal threshold current of the instability; we saw that the instability is aperiodic. The following question then arises: does the behaviour of the Pierce instability change if the beam current considerably exceeds the minimal Pierce threshold? An analysis [2.14, 2.15, 2.21, 3.1–3.3] led to the following new results. (For simplicity, these results are first considered here for the case of a planar system; at the end of this section, we will also analyse the cylindrical system.)

First, it is found that the Pierce instability remains aperiodic if

$$(2n - 1)\pi < \alpha < 2n\pi \tag{3.1}$$

where

$$\alpha = \omega_1 L / u_0 \tag{3.2}$$

is the dimensionless Langmuir frequency of the beam

$$\alpha^2 = 4\pi e j L^2 / m u_0^3 \tag{3.3}$$

and j is the beam current density. The threshold value α_c is given by the relation

$$\alpha_c = (2n - 1)\pi. \tag{3.4}$$

Relation (3.1) shows that the aperiodic Pierce instability exists only in specific intervals of beam current variation.

Second, we find that the Pierce system we consider here is unstable outside the intervals of aperiodic instability: a new (oscillatory) instability develops, not yet discussed above. We also find that those ranges of variation of α in which the oscillatory instability dominates are complementary with respect to the intervals of aperiodic instability; they meet the condition

$$2n\pi < \alpha < (2n+1)\pi \qquad (3.5)$$

or

$$\alpha > \alpha_c \qquad \text{where} \qquad \alpha_c = 2n\pi \qquad (3.6)$$

and $n = 1, 2, 3. \ldots$

It should be emphasized that the oscillatory instability exists in a system in which the compensating ions are again assumed to be at rest; this is a purely electron instability. The oscillation frequency is several times lower than the reciprocal time of flight of electrons and increases as $(\alpha - \alpha_c)$ increases.

In view of the oscillatory instability, the Pierce system treated here is unstable at almost any beam current. To put this statement in a more stringent form: on the scale of variation of the parameter α, very narrow 'stability islands' exist in the approximation chosen, that is, as $M \to \infty$. Using them, it is possible to trace an interesting evolution of the system, in its transformation from a stable state to a state of oscillatory instability.

Third, it is found that the increments (γ) of the aperiodic and oscillatory Pierce instabilities (close to their thresholds) are given by the relations

(*a*) for the aperiodic instability

$$\gamma = i\alpha_c\Delta/4 \qquad (3.7)$$

(*b*) for the oscillatory instability

$$\gamma = i\alpha_c\Delta/9. \qquad (3.8)$$

We also find that the oscillation frequency Ω is of the order of the reciprocal time of flight of electrons; to be more precise, it is

$$\Omega = (\alpha\Delta/3)^{1/2} \qquad (3.9)$$

where

$$\Delta = \alpha - \alpha_c. \qquad (3.10)$$

The quantities α_c for the two versions of the instability discussed above are given by (3.4) and (3.6).

The Pierce instability zones indicated here are illustrated in figure 3.1 which plots the results obtained by numerical simulation in a number of independent papers [2.14, 2.15, 2.21; 3.1–3.3].

Let us look in more detail at the characteristics of the oscillatory instability of the Pierce system under consideration [3.3]. First of all, we

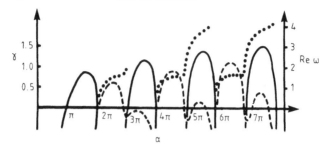

Figure 3.1   The Pierce instability increment and the frequency of electron oscillations, for ions at rest: increment and oscillation frequency of the Pierce instability as functions of beam current, $\alpha = \omega_1 L/u_0$. Solid curves plot the increment of aperiodic instability, dashed curves, that of oscillatory instability, and dotted curves give the oscillation frequency. Obviously, the zones of aperiodic and oscillatory instabilities mostly alternate and the oscillatory instability increment in the zones of aperiodic instability is either small ($\gamma > 0$) or negative. The quantities γ and $\operatorname{Re}\omega$ are given in units of $\pi u_0/L$.

notice the fact that the scale of the parameter α does reveal the stability 'islands' mentioned above, with ions at rest ($M \to \infty$). They are located close to the right-hand boundaries of the oscillatory instability zones; for example, the first of them lies between the points $\alpha = 3\pi$ and $\alpha = 2.9\pi$. If α is reduced below the left-hand boundary of the stability region, $\alpha = 2.9\pi$ (see figure 3.1), the system displays the following intriguing evolution [3.3, 2.21]. First to arise are regular oscillations at a frequency $\omega = \omega_1/6$. Then the first bifurcation of doubling the oscillation period occurs at $\alpha \simeq 2.8646\pi$: in addition to the frequency ω, its subharmonic $\omega/2$ also appears. Then, at $\alpha = 2.8592\pi$, the next bifurcation of doubling the oscillation period occurs, adding a subharmonic $\omega/4$ to the already present ω and $\omega/2$. The next bifurcations resulting in additional frequencies $\frac{1}{8}\omega$ and $\frac{1}{16}\omega$ occur at $\alpha = 2.85813\pi$ and $\alpha = 2.85792\pi$, respectively. Further steps of the evolution are practically unresolvable since the intervals of change in α decrease drastically. It is interesting to note that the distances between neighbouring bifurcations on the α scale differ by a factor of five, which is quite close to the value 4.6692, corresponding to the familiar Feigenbaum scenario of the origin of turbulence [3.4]. As α further decreases, namely, at $\alpha = 2.85786\pi$, a stochastic attractor arises; oscillations become almost random. Finally, when $\alpha = 2.8491\pi$, the so-called 'crisis' of stochastic attractor occurs and the system transforms to the state with a virtual cathode. These results are illustrated in figures 3.2–3.4, and are explained in the respective captions.

We see that a non-linear evolution of the Pierce system results both in the aperiodic instability mode ($\pi < \alpha < 2\pi$) and, within the predominant part of the oscillatory instability region ($2\pi < \alpha < 2.85\pi$), in the formation

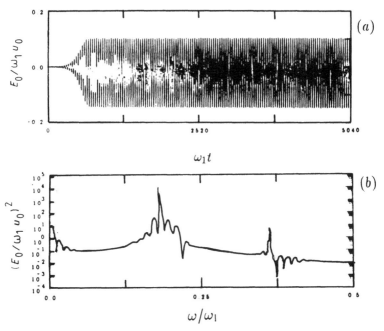

Figure 3.2 (a) Oscilloscope trace and (b) power spectrum of non-linear electron oscillations in the Pierce diode at $\alpha = 2.88\pi$. Oscillation amplitude is shown in arbitrary units.

of a virtual cathode which reflects some electrons back. No virtual cathode is formed within a small part of the oscillatory instability region ($2.85\pi < \alpha < 3\pi$): the depression of potential is limited by the level which is lower than the beam electron energy.

We have discussed two first instability zones. Similar phenomena occur in other zones, shifted by 2π on the α scale [3.3, 3.5].

The influence of the motion of compensating ions on the characteristics of the instability of the system under consideration is a problem of principal importance. An answer to this question was obtained in numerical simulations [2.14, 2.15, 2.21]. It is like this. First, as the parameter $\delta = m/M$ (characterizes the mobility of ions) increases from zero ($\delta = 0$ if $M \rightarrow \infty$), the dependences of instability increments and oscillation frequencies on the parameter α undergo severe deformation. Second, the stability islands disappear. Third, *new* oscillation branches appear. Fourth, the oscillatory instability increment explicitly exceeds the increment of aperiodic instability. This happens when the length of the system is greater than the Debye radius (1.1) of the beam electrons by a factor of at least $(M/m)^{1/2}$. This relation has a clear physical meaning: if the system's length is sufficiently large (compared with the Debye radius of beam electrons), the finite length of the system (which is the cause of the aperiodic instability)

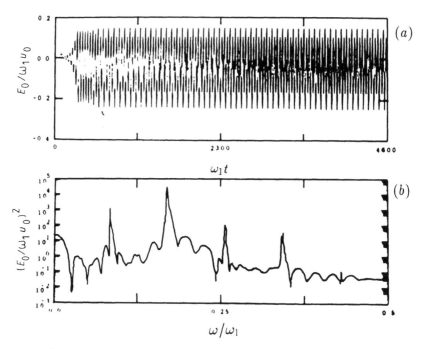

Figure 3.3   The same as in figure 3.2 but for $\alpha = 2.86\pi$. The oscillation period is seen to double.

plays a rather insignificant role. (Thus the aperiodic instability increment $\gamma \to 0$ as $L \to \infty$.) The main role is then played by the electron–ion oscillations which survive in an unbounded system. These so-called 'Buneman oscillations' are discussed in the next section.

In the case of cylindrical beam geometry, the qualitative relations outlined above are retained, with the following (fairly obvious) quantitative modifications: in expression (3.2) for α, the quantity ω_1 must be replaced by $(\omega_1^2 - k_\perp^2 u_0^2)^{1/2}$, where the quantity $k_\perp$ is defined by formula (2.14). In the relativistic case (cylindrical beam geometry), ω_1 is replaced by $(\omega_1^2/\gamma_0^3 - k_\perp^2 u_0^2)^{1/2}$.

3.2   Electron–Ion Instability of Quasineutral Electron Beams

In the preceding section, when analysing the Pierce system with a quasineutral electron beam of a finite (not too large) length, we assumed that the compensating ions were at rest ($M = \infty$). In this case the instability is caused by the boundary conditions, namely, by the positive feedback via the electrons of the external circuit. The elimination of boundaries by increasing the length of the system ($L \to \infty$) brought it

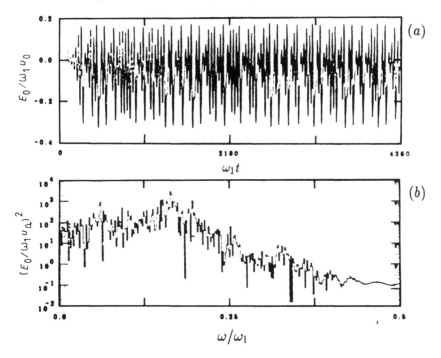

Figure 3.4 The same as in figure 3.2 but for $\alpha = 2.85\pi$. A 'strange' (stochastic) attractor is formed [3.3].

to a stable state. This is natural since if $L \to \infty$, the particles realizing the positive feedback were eliminated.

However, if ions compensating the beam are mobile ($M \neq \infty$), they can also play the role of positive feedback. In this case, an electron–ion instability arises, which we will now discuss.

Historically, this instability was first discovered by Pierce [1.4], then by Budker [3.6] and later by Buneman [3.7, 3.8]. For this reason, although this instability is known as the Buneman instability, it would be more correct to refer to it as the Pierce–Budker–Buneman instability. In order not to confuse it with the Pierce instability discussed above, we will use the term 'Budker–Buneman instability' (BB).

Let us consider a spatially uniform monoenergetic electron beam propagating through a 'background' of compensating *mobile* positive ions. The non-perturbed ion density n_+ equals the non-perturbed beam electron density n_1. For the sake of simplification, we assume the beam to be non-relativistic; the specifics of a relativistic beam will be considered later. We will analyse the stability of such systems by the dispersion equation method, which constitutes a relation between the frequency ω of possible oscillations of charged particles and the wave vector $\mathbf{k}$. The following arguments are employed to derive the dispersion equation. Let an electric

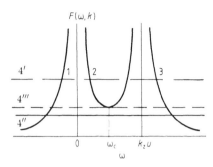

$$F(\omega, k)$$

Figure 3.5   Dispersion curves: 1–3, branches of the function
$y = F(\omega)$ for $k = $ constant; 4, lines $y = 1$; 4', stability; 4'',
instability; 4''', threshold (critical) regime.

field $E = -\,\mathrm{grad}\,\psi$, oscillating with time at a frequency ω, arise in a random manner in the system under consideration. For simplicity, we begin with the case of the oscillations of charged particles occurring only along E†. In this geometry, the field E generates oscillations of ions at a frequency ω and velocity $-eE/i\omega M$ (M is the ion mass) which correspond to the alternating density of the ionic current, $j_+ = -e^2 n_+ (i\omega M)^{-1} E$; we assume that the time dependence of all variables is described by $\exp i(kr - \omega t)$. According to the continuity equation $\partial \rho_+ / \partial t = -\,\mathrm{div}\,j_+$, the density of the ionic space charge is $\rho_+ = (e^2 n_+ / \omega^2 M)\,\mathrm{div}\,E$. The effect of the field E on the electrons of the beam is (in principle) the same but owing to the Doppler effect, the electrons (moving at a velocity u) are induced to oscillate at a frequency $\omega' = \omega - ku$ (so far, we neglect relativistic effects); hence, the electron space charge density is

$$\rho_- = \frac{e^2 n_1 \,\mathrm{div}\, E}{(\omega')^2 m} = \frac{e^2 n_1 \,\mathrm{div}\, E}{(\omega - ku)^2 m}.$$

As a result, Poisson's equation $\mathrm{div}\, E = 4\pi(\rho_+ + \rho_-)$ gives the following equation for the space-charge waves ($\mathrm{div}\, E \neq 0$):

$$\left.\begin{aligned}
\frac{\omega_1^2}{(\omega - ku)^2} + \frac{\omega_+^2}{\omega^2} &= 1 \\[2mm]
\frac{\omega_1^2}{(\omega')^2} + \frac{\omega_+^2}{\omega^2} &= 1
\end{aligned}\right\} \qquad (3.11)$$

where $\omega_1 = (4\pi n_1 e^2/m)^{1/2}$ and $\omega_+ = (4\pi n_+ e^2/M)^{1/2}$ are the Langmuir (plasma) frequencies for electrons and ions, respectively, and $n_1 = n_+$.

---

† This situation is realized either in the case of planar geometry or in the absence of magnetic field.

The solution of dispersion equation (3.11) (of degree four with respect to ω) is best obtained by grapho-analytic method. Denoting the left-hand part of the equation by $F(\omega)$, we plot the branches of the function $F(\omega)$ (see figure 3.5). We see that these branches cross the line $F(\omega) \equiv 1$ either at four or at two points, depending on the parameters of the dispersion equation. In the case of four intersection points, all four roots of the dispersion equation are real and the system is stable. In the second case, only two roots are real, and among the remaining two (complex conjugate) roots, one has a positive imaginary part: $\omega = \omega_r + i\gamma$, $\gamma > 0$; instability takes place in this situation: oscillations grow with time at an increment γ. In the critical mode corresponding to the instability threshold, the line $F(\omega) \equiv 1$ is tangent to the middle branch of the function $F(\omega)$, and $\partial F/\partial \omega = 0$. In this critical case, the oscillation frequency is

$$\omega_c = \frac{ku}{1 + (M/m)^{1/3}} \tag{3.12}$$

where $k = k_z$ is the wave number corresponding to the direction of beam propagation along the z axis and the subscript with frequency ω corresponds to the critical regime.

Substituting ω_c from (3.12) into (3.11), we obtain an expression for the threshold (critical) beam electron density n_{1c} after which the instability sets in:

$$\omega_{1c}^2 \equiv \frac{4\pi n_{1c} e^2}{m} = \frac{k^2 u^2}{[1 + (m/M)^{1/3}]^3}. \tag{3.13}$$

For the threshold beam current we have (for beam radius a)

$$I_c \equiv \pi a^2 e n_{1c} u = \frac{ma^2}{4e} \frac{k^2 u^3}{[1 + (m/M)^{1/3}]^3}. \tag{3.14}$$

The instability increment was found in Buneman's theoretical paper [3.7] (see also [3.8]). Figure 3.6 plots the increment γ and the frequency $\mathrm{Re}\,\omega$ as functions of the ratio $\omega_1^2/(ku)^2$. The values of γ and $\mathrm{Re}\,\omega$ are plotted in units of $\omega_1(m/M)^{1/3}$; we assumed $m/M = 1836$ (atomic hydrogen). The maximum of the increment is reached for

$$\omega_1 = ku \tag{3.15}$$

and is implied by the relation

$$\gamma_{\max} = \frac{\sqrt{3}}{2^{4/3}} (m/M)^{1/3} \omega_1. \tag{3.16}$$

Note that

$$\mathrm{Re}\,\omega = \frac{1}{2^{4/3}} (m/M)^{1/3} \omega_1. \tag{3.17}$$

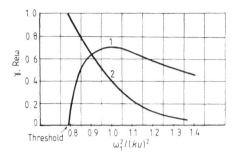

Figure 3.6   The parameters of the electron–ion instability. Curve 1, γ; curve 2, $\text{Re}\,\omega$, $\gamma_{max} = (\sqrt{3}/2^{4/3})\omega_1(m/M)^{1/3}$. If $\omega_1 \gg \omega_{1c}$ (far beyond the threshold), then $\gamma \simeq (m/M)^{1/2}ku \ll \gamma_{max}$.

Formulas (3.15)–(3.17) describe the so-called resonance instability regime. A certain deviation of the oscillation frequency (3.17) from (3.12) need not seem surprising: indeed, (3.12) corresponds to the instability threshold (when $\omega_1 = \omega_{1c} < ku$), while (3.17) corresponds to the resonance mode with finite overcriticality, with $\omega_1 = ku$. Figure 3.6 shows that as overcriticality ($\omega_1^2 - \omega_{1c}^2$) increases, the oscillation frequency decreases substantially. Thus the oscillation frequency is lower than the 'Buneman' value (3.17) by a factor of several tens already at the beam current only twice greater than the threshold (for hydrogen). (This important behaviour is rarely taken into account in the literature, however strange this may seem.)

If the overcriticality is large ($\omega_1 \gg ku$), the increment γ is easily found from (3.11):

$$\gamma \simeq (m/M)^{1/2}ku. \tag{3.18}$$

We see that this increment is appreciably lower than the resonance value (3.16).

The instability discussed in this section thus manifests itself in the generation of oscillations in which both beam electrons and ions participate. A spectacular property of this system is the fact that two such different components as beam electrons (with relatively high natural frequency ω_1) and heavy ions of the ambient medium (with much lower natural frequency $\omega_+ = (m/M)^{1/2}\omega_1 \ll \omega_1$) are in fairly efficient (resonant) interaction with each other.

In order to clarify the physical meaning of the results, we resort to the following line of reasoning. Assume first that ions are at rest ($M \to \infty$). Then dispersion equation (3.11) implies (for simplicity, we choose the case of planar geometry)

$$\frac{\omega}{k} = u \pm \frac{\omega_1}{k}. \tag{3.19}$$

This relation describes stationary waves of the space charge in the beam: the fast wave (for the plus sign) and the slow wave (for the minus sign). In the reference frame of the beam, these are simply the Langmuir oscillations at the electron natural frequency ω_1 propagating both in the direct and in the reverse directions at the phase velocity $\pm\omega_1/k$. In the laboratory reference frame, their phase velocity is shifted by the velocity of the beam electrons. Let us now take into account the mobility of ions. If they oscillated independently of electrons, the oscillation frequency would equal the natural ion frequency ω_+ and the phase velocity would be ω_+/k. If, however, both components of the system oscillate (both electrons and ions), the qualitative condition of the resonance will be very close to

$$u - \frac{\omega_1}{k} \simeq \frac{\omega_+}{k} \qquad ku - \omega_1 \simeq \omega_+. \tag{3.20}$$

This relation signifies that the slow space-charge wave propagates synchronously with the ion wave. This mechanism of generation of oscillations is also fundamental for travelling-wave tubes [1.6]: in this case, the role of the wave in the slowing-down structure (which subtracts energy from the space-charge wave of the beam) is played by the ion wave. The resonance condition (3.20) is satisfied only when beginning with a certain threshold of beam density: in order for the space-charge wave of the beam to be synchronous with the ion wave, its phase velocity $u - \omega_1/k$ must greatly decrease (in comparison with u). This is possible only for a sufficiently high value of ω_1, not differing considerably from the value of ku. This is the physical meaning of relations (3.13) and (3.14) which determine the threshold of the instability under consideration.

The analysis above makes it possible to arrive at another qualitative conclusion. Assume that the beam current exceeds the instability threshold. The amplitude of the space-charge waves then begins to grow. When it reaches a sufficiently high value at which the wave potential is

$$\varphi \geqslant \frac{m}{2e}\left(u - \frac{\omega}{k}\right)^2 \tag{3.21}$$

beam electrons get trapped into the wave. Since the phase velocity of the wave is much lower than the initial beam velocity, the average beam velocity decreases substantially. A detailed theoretical analysis combined with numerical simulation shows that the trapping of electrons into the wave at a sufficiently high beam current must—in an infinitely large planar system—trigger beam disruption, that is, the formation of the virtual cathode in the beam [3.9, 3.10, 2.10]. If this is correct, the limiting beam current dictated by this type of instability, (3.14), is shown to be quite close to the first threshold of the aperiodic Pierce instability.

According to numerical simulation [3.11], the non-linear evolution of the Buneman instability leads to chaos via 'intermittency'. (For an introduction to modern concepts of the physics of turbulence and on various scenarios of the onset of chaos, we recommend the excellent monograph of Manneville [3.12].)

Another side to this phenomenon is, as in the case of the Pierce instability discussed above, the acceleration of ions to high energies that can greatly exceed that of electrons in the beam [3.13–3.16].

Chapter 6 discusses to what extent these predictions of the theory are confirmed under the conditions of actual experiments (see sections 6.3 and 6.4).

Until now, we treated the case of planar geometry, in which all particles oscillate along the electric field of the oscillations. Let us turn now to a different case, which better suits the geometry of the experiment to be discussed below. In this more general case, the direction of oscillation of particles is at an angle θ to the direction of $\boldsymbol{E}$ (this angle can be different for electrons and ions). For instance, let electrons be magnetized by a strong field $\boldsymbol{H}$ pointing along the beam axis (z axis) at an angle θ to $\boldsymbol{E}$ while ions are not magnetized, which occurs if

$$\omega_{Hi} \ll \omega \ll \omega_{He} \tag{3.22}$$

where $\omega_{Hi} = eH/(Mc)$ and $\omega_{He} = eH/(mc)$ are the ionic and electronic Larmor frequencies, respectively. In this case, electrons oscillate along the z axis and ions oscillate along $\boldsymbol{E}$, that is, along the wave vector $\boldsymbol{k}$; also,

$$\cos \theta = \begin{cases} k_z/k & \text{for electrons} \\ 1 & \text{for ions.} \end{cases} \tag{3.23}$$

Turning now to deriving the dispersion equation (3.11), we find that in the case of (3.23), the terms on the left-hand side of (3.11) must be multiplied by $\cos^2 \theta$. Indeed, the wavelength of the oscillations we discuss is a function of the direction in space. Thus, the wavelength in the direction $\boldsymbol{H}_z$ is $\lambda = \lambda_z = 2\pi/k_z$. In the direction of oscillation of ions, however (i.e., in the direction of $\boldsymbol{E}$), we find $\lambda = 2\pi/k = \lambda_z \cos \theta < \lambda_z$. Therefore, the oscillations in the direction of $\boldsymbol{H}$ have the largest wavelength. This is the explanation of the physical meaning of the factor $\cos^2 \theta$ appearing in the sought dispersion equation of oscillations: the factor $\cos \theta$ appears once because the oscillations of magnetized particles are caused by the force $eE \cos \theta$, so that the oscillatory electron current $j_- \propto \cos \theta$. It appears again because owing to the continuity equation, the space charge of these particles is proportional to the derivative of the current density j_- along the direction $\boldsymbol{H}_z$ which is less by a factor $\lambda_z/\lambda = k/k_z$ than the derivative in the direction of $\boldsymbol{E}$. As a result, the amplitude ρ_1 of space-charge density oscillations of electrons is proportional to the ratio k_z^2/k^2. Under the conditions of an actual experiment (when standing waves with longitudinal wave number

$k_z = n\pi/L$ are formed), the quantity k_z^2/k^2 may be of the order of the ratio of the beam length squared to beam radius squared, and it is essential that this factor be taken into account.

In the case at hand, that is, for (3.23), the dispersion equation takes the form not of (3.11) but

$$\frac{\omega_1^2}{(\omega - k_z u)^2} \frac{k_z^2}{k^2} + \frac{\omega_+^2}{\omega^2} = 1. \tag{3.24}$$

By analogy to the arguments above, we find the oscillation frequency at the instability threshold

$$\omega_c = \frac{k_z u}{1 + ((M/m) k_z^2 k^{-2})^{1/3}} \tag{3.25}$$

the threshold (critical) electron beam density (n_{1c})

$$\omega_{1c}^2 = \frac{4\pi n_{1c} e^2}{m} = \frac{k^2 u^2}{\left[1 + ((m/M) k^2 k_z^{-2})^{1/3}\right]^3} \tag{3.26}$$

and the threshold beam current (for beam radius a)

$$I_c = \pi a^2 e n_{1c} u = \frac{ma^2}{4e} \frac{k^2 u^3}{\left[1 + (\frac{m}{M} k^2 k_z^{-2})^{1/3}\right]^3}. \tag{3.27}$$

From the experimental standpoint, the case of maximum interest is that of a beam of radius a propagating along the axis of a metal housing of radius $R \gg a$ (the beam length $L \gg R$). The threshold instability current is then dictated by the quantity $k^2 = k_z^2 + k_r^2$, where k_r of the largest (and therefore the most important) oscillation mode is, by virtue of (2.14),

$$k_r^2 \simeq \frac{2}{a^2 \ln(R/a)} \tag{3.28}$$

and the minimum value of k_z (again for the largest mode) can be set equal to π/L. The expression (3.27) is equivalent to the instability condition

$$\left(\omega_+^2 \frac{k^2}{k_z^2}\right)^{1/3} + (\omega_1^2)^{1/3} > (k^2 u^2)^{1/3}. \tag{3.29}$$

In this particular case, the instability increment is determined not by the value of ω_1 (as it is in (3.16)) but by the 'reduced' Langmuir frequency $\omega_1 k_z/k \ll \omega_1$, where the value of $k \simeq k_r$ is found from (3.28).

Returning to the possibility of beam current disruption due to the Budker–Buneman instability, we notice that this possibility has so far been demonstrated only for the resonant regime of instability, in which $k_z \gg k_r$. As for the non-resonant regime $(k_z \ll k_r)$, no such conclusions were derived. This fact must be borne in mind when analysing the experimental data discussed in chapter 6. In connection with the aspects outlined above, it is very advisable to pay close attention to an original paper [3.17].

3.3   Electron–Electron Instability

Let us consider an electron beam propagating in a relatively dense plasma with electron concentration n_2 and the Langmuir frequency $\omega_p = (4\pi n_2 e^2/m)^{1/2}$. The dispersion equation of electron oscillations will be similar to the equation of electron–ion oscillations, with the ion frequency ω_+ replaced by ω_p in the one-dimensional case and by $\omega_p k_z/k$ in the case of cylindrical geometry with magnetized electrons. In this last example, (3.24) implies that

$$\frac{\omega_1^2 k_z^2/k^2}{(\omega - k_z u)^2} + \frac{\omega_p^2 k_z^2/k^2}{\omega^2} = 1. \tag{3.30}$$

By analogy to what we did in section 3.2, we find relations for the instability threshold, oscillation frequency in the vicinity of the threshold and the maximum increment:

$$(\omega_{1c}^2)^{1/3} + (\omega_p^2)^{1/3} = (k^2 u^2)^{1/3} \tag{3.31}$$

$$I_c = \frac{ma^2}{4e}\,\omega_{1c}^2 u = \frac{ma^2}{4e}\,k^2 u^3 \left[1 + \left(\frac{n_2}{n_1}\right)^{1/3}\right]^{-3} \tag{3.32}$$

$$\omega = k_z u \left[1 + \left(\frac{n_1}{n_2}\right)^{1/3}\right]^{-1} \lesssim \omega_p \tag{3.33}$$

$$\gamma = \frac{\sqrt{3}}{2^{4/3}} \left(\frac{n_1}{n_2}\right)^{1/3} \frac{k_z}{k}\,\omega_p. \tag{3.34}$$

Equation (3.31) leads to interesting conclusions: if

$$\omega_p > ku \tag{3.35}$$

then the instability sets in at any beam density, and if

$$\omega_1 > ku \tag{3.36}$$

then the instability sets in at any plasma density. (Obviously, however, the instability increment depends both on the beam density and on the plasma density.)

By analogy to (3.20), we can write a qualitative condition of instability (for the one-dimensional case) in the form

$$u - \omega_1/k \simeq \omega_p/k. \tag{3.37}$$

This instability was historically the first experimentally detected beam instability: this was the instability which was responsible for the 'Langmuir paradox' (see chapter 1). As we have already shown, this instability is based on the effect of quasiresonant interaction between a charged particle beam and a wave whose phase velocity is 'slightly' lower than the beam particle velocity. This effect, causing the beam instability, differs from the familiar *spontaneous* Vavilov–Čerenkov emission of radiation by an individual charged particle in that the former is *collective* (or *induced*), not spontaneous [3.10, 3.18]. The point is that if the particle beam density is uniform, the total effect of all events of spontaneous emission by individual particles cancels out, because the phases of the emitted waves are random relative to the particles of the beam. This effect arises only if the emission of radiation is *induced* by nature: the radiation field modulates the beam density and causes the particle bunches to be concentrated into the deceleration phases of the field. However, this phasing may be imperfect, so that the contributions of individual particles to the total electric field of the instability may become non-coherent. For all particles to move in coherence, it is necessary that they all satisfy the condition

$$u_0 > \omega/k \tag{3.38}$$

that is, the particle velocity must exceed the phase velocity of the wave. This condition can be derived [1.16] from the law of conservation of the energy of a beam particle which propagates synchronously with the electrostatic wave whose potential, φ, and the electric field, $E = -\mathrm{d}\varphi/\mathrm{d}x$ along the direction x of beam propagation, obey the equation

$$\frac{m}{2}\left(u - \frac{\omega}{k}\right)^2 - e\varphi = \text{constant} .$$

Hence, the wave increases the particle velocity by

$$\delta u = \frac{e\varphi}{m}/(u_0 - \omega/k).$$

We see that for bunching the particles in deceleration phases of a wave field, that is, for the wave pumping by the beam, condition (3.38) must be satisfied. However, this condition is necessary but not sufficient: if only this condition is satisfied, the phasing of beam bunches will take place only in the region of such wave phases where the field is almost zero, that is, there is no energy exchange between the beam and the wave [3.18]. In order that the equilibrium phase of the field be essentially non-zero, it is necessary (and sufficient) to satisfy another condition: the beam current must be greater than the instability threshold (in this particular case, relation (3.13) holds). Using the instabilities outlined above as examples, we readily see that if the threshold condition is met, condition (3.38) holds automatically. This occurs because the phase velocity of the wave decreases as the beam density increases (see relation (3.33)). It is this *collective* effect that determines the necessary lag of the wave with relation to the beam (see (3.38)).

If both condition (3.38) and the condition for the current to be greater than the threshold are satisfied, the instability is of hydrodynamic nature. This instability regime is sometimes called the *collective induced Vavilov–Čerenkov effect*. I think that a more appropriate name would be the *anomalous Doppler effect*. This effect will be discussed in chapter 4.

The arguments given above clarify the physical meaning of the hydrodynamic beam instability in which *all* particles of the beam pump the wave field coherently. The hydrodynamic mechanism of beam instability is realized in beams with small velocity spread. In beams with large velocity spread, the kinetic mechanism of beam instability operates, in which the greater part of the beam pumps the wave while the smaller part absorbs it (the inverse Landau damping effect). This is a less efficient mechanism (for details, see chapter 4).

The beam instability determines the beam relaxation in the velocity space (the 'Langmuir paradox'). This process is determined by a wide range of non-linear collective effects which include other types of instability. Thus, we find among them the modulation instability which results in soliton generation (chapter 10). An analysis of these phenomena, that is, a detailed discussion of the interpretation of the 'Langmuir paradox' would lead us far beyond the scope of this book. For this reason, I will only cite the relevant special literature: [1.16] and [3.19].

3.4   Beam–Drift Instability of a Spatially Non-uniform Beam–Plasma System in a Magnetic Field

Until now, we have discussed the instability of a spatially uniform beam–plasma system with respect to the pumping up of axially symmetric oscillations. If the particle density distribution is uniform, the generation of oscillations having no axial symmetry (in which the wave vector has three components: $k^2 = k_z^2 + k_r^2 + k_\varphi^2$) would require a higher threshold current proportional to k^2. This increase in the threshold would be quite considerable since the radial and the azimuthal wave numbers (k_r and k_φ) are typically of the same magnitude ($\propto 1/a$) and are much greater in long beams than the longitudinal wave number k_z.

However, the situation changes drastically if the beam density distribution in the cross section of the electron beam is essentially non-uniform, that is, if $\nabla n_1 \neq 0$. In this case, a new mechanism of instability arises, connected with the drift motion of particles in crossed fields: azimuthal electric field $\boldsymbol{E}_\varphi$ of oscillations and the longitudinal magnetic field $\boldsymbol{H}_z$. This is essentially the same mechanism as the one which causes the drift, or gradient, instability (also known as *universal instability*) of non-uniform plasma which is of great interest in controlled nuclear fusion research. The difference lies only in the form of the velocity distribution function of electrons, which is not Maxwellian in this particular case, but also involves the particle beam. Therefore, the instability of a strongly non-uniform beam in a plasma, which manifests itself in pumping up axially non-symmetric oscillations, combines features of an 'ordinary' beam instability in a uniform plasma and a drift instability of a non-uniform plasma. We will refer to it as the *beam–drift instability* (see also [3.20, 3.21 and 2.13]). We will now derive the dispersion equation for oscillations in a beam–plasma system having no axial symmetry. We will consider the more interesting case in which the electrons of the beam and the plasma are magnetized while the ions are not; to be precise, we will consider electron–ion oscillations in the frequency range

$$\omega_{Hi} < \omega < \omega_{He} \qquad k_z u. \tag{3.39}$$

We ignore the thermal motion of particles. In this case, the only new effect due to the expected presence of the perturbed azimuthal field E_φ and the radial gradient of charged particle density is the radial drift of beam and plasma electrons at a velocity equal to eE_φ/H_z. As a result of this drift, radial electron currents are generated: the current density is $j_\perp \simeq in_2 eck_\varphi \psi/H$ where n_2 is the particle density and ψ is the perturbed potential and $E_\varphi = -ik_\varphi \psi$. These currents change the amplitude of oscillations of the space charge of particles, in accordance with the continuity equation

$$\frac{\partial \rho}{\partial t} = -\operatorname{div}(j_\parallel + j_\perp)$$

where $j_\parallel$ is the density of the longitudinal (along $\boldsymbol{H}$) oscillatory current of particles due to the oscillation field E_z, that is, $j_\parallel = -(ne^2/i\omega m)E_z$. For instance, for the plasma electrons we find

$$\operatorname{div} j_\parallel = -e^2 n_2 \frac{\partial E_z}{\partial z} \frac{1}{i\omega m} = -n_2 e^2 k_z^2 \psi \frac{1}{i\omega m}$$

$$\operatorname{div} j_\perp \simeq \frac{-ecE_\varphi}{H} \frac{\partial n_2}{\partial r} \simeq \frac{n_2 e^2 k_\varphi}{i\omega_H m} \frac{\psi}{R} \simeq \frac{n_2 e^2 S \psi}{i\omega_H m R^2}$$

where R is the characteristic transverse scale (the 'plasma radius'); S is azimuthal mode number, that is, the length of the perimeter of the cross section of the system in units of azimuthal wavelength; and $k_\varphi \simeq S/R$. Therefore, the alternating space-charge density of plasma electrons is

$$\rho_e \simeq e^2 n_2 k_z^2 \frac{\psi}{m\omega^2} - e^2 n_2 S \frac{\psi}{R^2 m\omega\omega_H}.$$

The expression for the alternating space-charge density ρ of beam electrons must be quite analogous, with the replacements $n_2 \rightarrow n_1$, $\omega \rightarrow (\omega - ku)$ (the Doppler effect), and $R^2 \rightarrow a^2$ where a is the beam 'radius':

$$\rho_1 \simeq e^2 n_1 k_z^2 \frac{\psi}{m(\omega - k_z u)^2} - e^2 n_1 S \frac{\psi}{a^2 m(\omega - k_z u)\omega_H}.$$

Finally, the variable ion space-charge density ρ_+ is obviously given by the expression

$$\rho_+ \simeq e^2 n_+ k^2 \frac{\psi}{M\omega^2}$$

since ions oscillate along the resulting electric field $E = -\operatorname{grad}\psi$. Substituting the expressions ρ_1, ρ_2, ρ_+ into Poisson's equation $\nabla^2\psi = -k^2\psi = -4\pi(\rho_+ + \rho_e + \rho_1)$, we obtain a dispersion equation of the type $F(\omega, k) = 1$, with five terms in the left-hand side—one ionic term and two pairs of electronic ones:

$$\frac{\omega_1^2 k_z^2/k^2}{(\omega - k_z u)^2} - \frac{2S\omega_1^2}{a^2 k^2 \omega_H (\omega - k_z u)} + \frac{\omega_p^2 k_z^2/k^2}{\omega^2} - \frac{2S\omega_p^2}{k^2 R^2 \omega_H \omega} + \frac{\omega_+^2}{\omega^2} = 1.$$

$$(3.40)$$

This equation differs from equations (3.24) and (3.30) in that it contains two 'drift' terms (the second and the fourth) connected with the drift motion of the beam and plasma electrons in crossed fields: the perturbed electric field E_φ and the main (longitudinal) magnetic field H_z. These additional terms, which do not vanish only if simultaneously $k_\varphi \neq 0$ and $\operatorname{grad}(n_1, n_2) \neq 0$, essentially modify the stability criterion of a plasma–beam system†. Let us now analyse dispersion equation (3.40). Turning to figure 3.5, we easily notice that other conditions being equal, the beam drift term (if $\omega < k_z u$, the term is positive) helps the middle branch of the function $F(\omega)$ to break away from the horizontal line $F = 1$. The final result should be a lowering of the instability threshold, unless this effect is cancelled out by another effect, already mentioned above: an increase in the instability threshold in response to an increase in k^2. Likewise, it is easy to see that the negative plasma drift term is conducive to the stabilization of the instability.

The presence of drift terms of first order in ω in the dispersion equation excludes any possibility of exact analytic determination of the critical (threshold) instability parameters by the method used in section 3.2. Correspondingly, we will use a different approach: we assume that the inequality

$$\omega < k_z u$$

is sufficiently strong. Neglecting ω in (3.40) in comparison with $k_z u$, we obtain the following effective instability threshold for the quasineutral

---

† The coefficient 2 appears in the drift terms because of a more detailed analysis of the effect of the specific radial distributions of the beam and plasma densities. A parabolic distribution was chosen above:

$$n_1(r) = n_1(0)(1 - r^2/a^2) \quad n_2(r) = n_2(0)(1 - r^2/R^2).$$

electron beam $(n_2 \ll n_1, n_+ \simeq n_1)$:

$$\omega_{1c}^2 = \frac{k^2 u^2}{1 + 2Su/a^2\omega_H k_z} \tag{3.41}$$

(if $\omega_1 > \omega_{1k}$, the value of ω^2 implied by (3.40) becomes negative, which corresponds to pumping up of oscillations). Correspondingly, the critical beam current which is physically equivalent to the instability threshold is equal to

$$I_c = \frac{ma^2}{4e} \frac{k^2 u^3}{1 + 2Su/a^2\omega_H k_z}. \tag{3.42}$$

If the second term in the denominators of (3.41) and (3.42) is large in comparison with unity

$$\frac{Su}{a^2\omega_H k_z} \gg 1 \tag{3.43}$$

(the case of strongly non-uniform small-radius beams in weak magnetic fields), then the drift effects discussed above considerably lower the instability threshold. To clarify the physical meaning of condition (3.43), assume that we are not dealing with an electron beam in a plasma but with a Maxwellian plasma with electron temperature $T_e = mv^2$, where v is the thermal velocity of electrons. Imagine a perturbation in this plasma, with the azimuthal component of the electric field $E_\varphi \simeq T_e/ea$. In this field, which is perpendicular to the longitudinal magnetic field H, electrons begin to drift along radii at a velocity $cE_\varphi/H = cT_e/eHa$. They cover the characteristic distance $a \simeq 1/k_r$ in a time eHa^2/cmv^2. This time will be less than the time of displacement of electrons by a distance $1/k_z$ of the order of the longitudinal wavelength of oscillations, provided $v/a^2\omega_H k_z > 1$. This condition means that the drift motion of particles perpendicular to H in the perturbed azimuthal electric field is fast in comparison with their motion in the perturbed longitudinal electric field. However, this condition differs from (3.43) only in that instead of the beam electron velocity u, it contains the thermal velocity v. We see that these conditions are physically similar. Therefore, (3.43) signifies that the radial drift motion of electrons in the perturbed azimuthal electric field crossed with the longitudinal magnetic field is more important than longitudinal motion of electrons in the perturbed longitudinal electric field. In other words, condition (3.43) signifies strong influence of drift effects on the 'ordinary' beam instability.

Drift effects mostly change the dependence of the instability threshold on the basic parameters of the system: if (3.43) holds, threshold (3.42) is found to be proportional to the magnetic field strength, electron velocity squared and beam radius squared (since k^2a^2 is practically independent of the beam radius). At the same time, (3.27) implies that the excitation threshold of axially symmetric electron–ion oscillations is independent of the magnetic field strength, is proportional to the cube of the electron velocity in the beam and is independent of its radius. Finally, we note that in contrast to the 'ordinary' electron–electron beam instability in a spatially uniform beam–plasma system, the beam–drift instability is rather insensitive to the velocity spread of beam particles; for instance, it can evolve in the 'plateau'-type regime or may even display a two-sided plateau (see [3.20]).

These differences play an important role in the experimental identification of instabilities (see chapter 6). Turning again to the similarity and differences between the beam–drift and the drift instabilities, we immediately notice the following. On one hand, when (3.43) is satisfied, the physical foundations of these two instabilities are principally identical. On the other hand, one consequence of the difference between the Maxwellian electron distribution function (plasma without a beam) and the $\delta-$ function (a quasineutral beam) is that although the instability threshold (in particle density) is zero in the former case, it is not only non-zero in the latter case but—according to (3.42)—is a very steep function of the main parameters of the system. This feature is found to be of principal importance for the experiment: the drift instability can be identified by considering only the dispersion properties while the beam–drift instability is also characterized by clearly pronounced and diverse threshold behaviour implied by (3.42).

If the beam current is considerably higher than the critical value, the beam–drift instability increment is of the order

$$\gamma \simeq (\omega_{Hi} k_z u/S)^{1/2} \tag{3.44}$$

or, if condition (3.43) is met,

$$\gamma \simeq \omega_+. \tag{3.45}$$

The relation of the beam–drift instability to the limiting current of the quasineutral electron beam and of a three-component system (in which, e.g., $n_2 \gg n_1$) is discussed in chapter 6.

3.5   Thresholds of Electron–Ion Instabilities of Relativistic Electron Beams

Effects of non-potentiality of oscillations may in general prove to be essential in the dynamics of relativistic beams, so that the potential approximation we chose may not be strictly valid. It is found that the effects of non-potentiality of oscillations leave the Pierce and Budker–Buneman instability thresholds almost unaffected. However, the influence of these effects on the threshold of the beam–drift instability needs to be taken into account [3.21, 3.22, 2.10].

Relativistic effects influence the wave dispersion and, hence, instability thresholds. Indeed, since the oscillation frequency in the beam reference frame is

$$\omega' = (\omega - ku)\gamma_0 \tag{3.46}$$

where

$$\gamma_0 = (1 - u^2/c^2)^{-1/2} \tag{3.47}$$

then the contribution F_e of the beam electrons to the dispersion equation of one-dimensional oscillations is

$$F_e = \frac{(\omega_1')^2}{\gamma_0^2(\omega - ku)^2} \qquad \text{where} \qquad (\omega_1')^2 = \frac{4\pi n_1' e^2}{m}$$

n_1' is the electron density in the beam reference frame and m is the electron mass. Since $n_1' = n_1/\gamma_0$, we have $F_e = (\omega_1^2)\gamma_0^{-3}/(\omega - ku)^2$: beam electrons in oscillations get 'heavier' by a factor of γ_0^3. Correspondingly, the dispersion equation for electron–ion oscillations in a spatially uniform beam takes the form

$$\frac{\omega_1^2}{\gamma_0^3(\omega - ku)^2}\frac{k_z^2}{k^2} + \frac{\omega_+^2}{\omega^2} = 1. \tag{3.48}$$

Here we assume, as we did before, that beam electrons are magnetized by the strong longitudinal magnetic field which satisfies condition (2.32) and that the intrinsic magnetic field of the beam is weak in comparison with the external longitudinal field.

The threshold of the Budker–Buneman electron–ion instability now changes to

$$I_c \equiv \frac{ma^2}{4e}\omega_{1c}^2 u = \frac{ma^2}{4e}\frac{k^2 u^3 \gamma_0^3}{\left[1 + ((m/M)(k^2/k_z^2)\gamma_0^3)^{1/3}\right]^3}. \tag{3.49}$$

It is of interest to remark that if the second term in brackets in the denominator is much greater than unity, then as $u \to c$, the current I_c ceases to be a function of the beam electron energy:

$$I_c \to \frac{Ma^2}{4e}k_z^2 c^3. \tag{3.50}$$

For instance, if ions are argon ions, $a = 1$ cm, $k_z \simeq \pi/L \simeq 3 \times 10^{-2}$ cm^{-1} (the beam length $L \simeq 100$ cm), then (3.50) gives $I_c \simeq 3 \times 10^5$ A.

Changing now to a spatially non-uniform system, we first check what the threshold of the beam–drift instability would be in the potential approximation. In this case, it is not difficult to show that only the first term changes in the dispersion equation which describes the beam–drift instability (see section 3.4); namely, the already familiar factor γ_0^3 will appear in the denominator of this term, so that the effective instability threshold becomes

$$I_c = \frac{ma^2}{4e} \frac{k^2 u^3 \gamma_0^3}{1 + 2Su\gamma_0^3/a^2\omega_H k_z} \tag{3.51}$$

where $\omega_H = eH/mc$. We see that in the relativistic regime, the drift effects influence in a decisive way the beam instability already when

$$\frac{Su\gamma_0^3}{a^2\omega_H k_z} > 1. \tag{3.52}$$

If the second term in the denominator of (3.51) is greater than unity, then as $u \to c$, the critical beam current is independent of the electron energy:

$$I_c \simeq \frac{mc^2}{4e} \frac{\pi a^2 \omega_H}{L} \tag{3.53}$$

where we have assumed $k_z = \pi/L$ and $k^2 \simeq k_r^2 + k_\varphi^2 \simeq 2/a^2$. For example, if $H = 10^4$ Oe, $a = 1$ cm and $L = 100$ cm, formula (3.53) gives $I_c \lesssim 300$ A.

If we now take into account the effects of non-potentiality [2.13, 3.22], we find that

$$I_c \simeq \frac{mc^2}{4e} \frac{\pi a^2 \omega_H}{L} \frac{\gamma_0^2}{2} \tag{3.54}$$

that is, the instability threshold grows by a factor of $\gamma_0^2/2$.

It is interesting to compare threshold (3.54) with the critical current (3.50) for the excitation of axially symmetric oscillations. We find that the former threshold is lower than the latter if

$$\frac{eH}{Mc} \equiv \omega_{Hi} < 2\pi c/L\gamma_0^2.$$

If γ_0 is not too large, this condition can be violated only in very strong magnetic fields. For example, the beam–drift instability threshold of argon for $\gamma_0 = 3$ and $L \simeq 10^2$ cm is lower than the Budker–Buneman instability threshold at $H < 10^6$ Oe. Therefore, the electron–ion instability of a magnetized relativistic beam must primarily be the beam–drift instability.

3.6   On Slipping-Stream Instabilities

Until now, we have discussed beam instabilities in a system which included, as a minimum, *two* components: primary beam electrons and particles of the 'medium' (ions, plasma electrons or electrons of the external circuit). In this case, the positive feedback (the foundation of any instability) operates through the particles of the medium. In this sense, the specific phenomena are the instabilities of electron beams with non-uniform velocity profile: the so-called instabilities of slipping streams. The instability may be caused by the inhomogeneity of the velocity profile of the movement along the external magnetic field. In this case, it is known as the slipping-stream instability. The instability may also be caused by an inhomogeneous velocity profile of rotation of particles in a (hollow) beam with non-compensated space charge (the rotation takes place in the electric field of the beam, crossed with the external magnetic field); in this case, the term *diocotron instability* is used. The mechanism of these instabilities has much in common with the mechanism of instability of sheared flows in hydrodynamics. For instance, this is the Kelvin–Helmholtz instability in the case of a velocity jump (in fact, this instability had been studied before the electron was discovered). The diocotron instability is connected with satisfying the Rayleigh criterion; this criterion is especially popular in the case of planar geometry—it then requires an inflection point on the flow velocity profile. In cylindrical geometry, the Rayleigh criterion is written in a somewhat more complicated form. In diocotron instability, this criterion predicts that the unstable characteristic is the velocity profile of the drift motion of particles in the magnetic field crossed with the electric field of the space charge of a hollow beam.

The slipping-stream instabilities are analysed in detail both in papers on hydrodynamics and plasma physics [3.23–3.27] and in special monographs on the physics of relativistic beams (e.g., see [3.28, 3.29]). Correspondingly, I only mention them here: both from the standpoint of their experimental observation (section 6.10) and in the light of the concept of negative-energy waves treated in the next chapter.

References

[3.1]   Cary J R and Lemons D S 1982 *J. Appl. Phys.* **53** 3303

[3.2]   Jungwirth K 1985 *Czech. J. Phys.* **35** 844

[3.3]   Godfrey B B 1987 *Phys. Fluids* **30** 1553

[3.4]   Feigenbaum M J 1979 *J. Stat. Phys.* **21** 669

[3.5]   Hörhager M and Kuhn S 1990 *Phys. Fluids* B2 2741

[3.6]   Budker G I 1956 *Atomnaya Energiya* **5** 9

[3.7]   Buneman O 1958 *Phys. Rev. Lett.* **1** 8

[3.8]   Buneman O 1959 *Phys. Rev.* **115** 503

[3.9]   Ishihara O, Hirose A and Langdon A B 1981 *Phys. Fluids* **24** 452, 610

[3.10] Kuzelev M V and Rukhadze A A 1987 *Uspekhi Fiz. Nauk* **152** 285 (Engl. Transl. 1987 *Sov. Phys.-USPEKHI* **30** 518)

[3.11]   Hamamatsu N 1990 *Phys. Fluids* **B2** 1780

[3.12]   Manneville P 1991 *Dissipative Structures and Weak Turbulence* (Boston: Academic)

[3.13]   Belova N G, Galeev A A, Sagdeev R Z and Sigov Yu S 1980 *Pis'ma v Zh. Eksp. Teor. Fiz.* **31** 551

[3.14]   Galeev A A, Sagdeev R Z, Shapiro V D and Shevchenko V I 1981 *Pis'ma v Zh. Eksp. Teor. Fiz.* **81** 572

[3.15]   Volokitin A S and Krasnoselskikh V V 1982 *Fizika Plazmy* **8** 800 (Engl. Transl. 1982 *Sov. J. Plasma Phys.* **8** 454)

[3.16]   Bulanov S V and Sasorov P V 1986 *Fizika Plazmy* **12** 54 (Engl. Transl. 1986 *Sov. J. Plasma Phys.* **12** 29)

[3.17]   Trubnikov B A and Zhdanov S K 1987 *Phys. Repts* **154** 201

[3.18]   Kalmykova S S and Kurilko V I 1988 *Uspekhi Fiz. Nauk* **155** 681 (Engl. Transl. 1988 *Sov. Phys.-USPEKHI* **31** 750)

[3.19]   Mishin E V, Ruzhin Yu Ya and Telegin V A 1989 *Interaction of Electron Flows with Ionospheric Plasma* (Leningrad: Gosmeteoizdat) (in Russian). See also references therein

[3.20]   Mikhailovsky A B 1975 *Theory of Plasma Instabilities* vols 1, 2 (Moscow: Atomizdat) (in Russian)

[3.21]   Rukhadze A A, Bogdankevich L S, Rosinsky S E and Rukhlin V G 1980 *Physics of High-Current Relativistic Electron Beams* (Moscow: Atomizdat) (in Russian)

[3.22]   Karbushev N I, Rukhadze A A and Udovichenko S Yu 1984 *Fizika Plazmy* **10** 268 (Engl. Transl. 1984 *Sov. J. Plasma Phys.* **10** 156)

[3.23]   Buneman O, Levy R H and Linson L M 1966 *J. Appl. Phys.* **37** 3203

[3.24]   Timofeev A V 1989 *Voprosy Teorii Plazmy* vol 17 ed B B Kadomtsev (Moscow: Energoatomizdat) p 157 (in Russian)

[3.25]   Leiman V G 1987 *Fizika Plazmy* **13** 1216

[3.26]   Leiman V G and Ovsyannikova O B 1989 *Fizika Plazmy* **15** 625

[3.27]   Leiman V G 1985 *Fizika Plazmy* **11** 563

[3.28]   Miller R B 1982 *An Introduction to the Physics of Intense Charged Particle Beams* (New York: Plenum)

[3.29]   Davidson R C 1974 *Theory of Non-neutral Plasmas* (Reading, MA: W A Benjamin)

4   Beam Instabilities as a Result of Active Coupling between Waves of Different Signs of Energy (Theory)

4.1   Waves of Positive and Negative Energy in Charged Particle Beams

One of the main results of chapter 3 can now be summarized as follows. Any 'hydrodynamic' instability of a charged particle beam occurs because the slow wave of the beam space charge is synchronous with the wave excited in the medium, pumps this medium wave and thereby is enhanced itself; obviously, the wave energy is drawn from the energy of the beam. For instance, the synchronism condition in the case of the Budker–Buneman instability signifies (roughly) that $u - \omega_1/k \simeq \omega_+/k$, and in the case of the pumping of Langmuir waves, that $u - \omega_1/k \simeq \omega_p/k$, which agrees with the theory outlined above. This picture has one spectacular feature which I will now discuss in detail: it is found that the slow beam wave grows precisely because it loses energy by exciting a wave in the medium. The explanation of this apparent paradox lies in realizing, as will be shown below, that the slow wave of the beam space charge carries, in contrast to the fast wave, a negative energy. By the definition of this concept, this means that the total energy (kinetic and electric) of the beam in which the wave has been excited is less than the kinetic energy of the same beam without the wave, and that this energy deficiency is the result of energy transfer from the slow beam wave to the wave carrying positive energy.

Assume that a 'monochromatic' wave (also specified to be plane, for the sake of simplification) propagates along a direction z in a homogeneous medium with dielectric permittivity ε; the electric field of the wave varies as $\exp i(kz - \omega t)$ (here and throughout the book, only electrostatic waves will be considered). If ε is independent of ω, that is, if there is no dispersion, then the wave energy density is given by the expression $W = \varepsilon E^2/8\pi$

where the overbar denotes averaging over one oscillation period and W is the sum of the energy of the electric field and of the kinetic energy of particle oscillations. If dispersion is non-zero, the expression for W is [4.1]

$$W = \frac{\overline{E^2}}{8\pi} \frac{\partial}{\partial \omega}(\varepsilon \omega) \tag{4.1}$$

where ε and ω stand for the real parts. If the medium is in thermodynamic equilibrium, the wave energy is always positive. For instance, the energy of Langmuir electron–plasma oscillations ($\omega = \omega_p$) is positive; these oscillations are implied by the dispersion equation $\varepsilon = 0$ where

$$\varepsilon = 1 - \omega_p^2/\omega^2. \tag{4.2}$$

However, the energy W in a non-equilibrium medium may be either positive or negative, depending on the type of dispersion. If $\partial(\varepsilon\omega)/\partial\omega < 0$, we will say that the wave carries negative energy [4.2]. For example, consider a quasineutral monoenergetic one-dimensional electron beam which propagates through a 'background' of compensating ions at rest. In this case, the longitudinal waves correspond to the dielectric permittivity

$$\varepsilon = 1 - \frac{\omega_1^2}{(\omega - ku)^2}. \tag{4.3}$$

By virtue of definition (4.1), the energy density of these waves is

$$W = \frac{\overline{E^2}}{4\pi} \frac{\omega \omega_1^2}{(\omega - ku)^3}. \tag{4.4}$$

We see that

$$\left. \begin{array}{l} W > 0 \text{ if } u < \omega/k \\ W < 0 \text{ if } u > \omega/k \end{array} \right\}. \tag{4.5}$$

The quantity $(\omega - ku)$ is found from the dispersion equation $\varepsilon = 0$, whence

$$\omega - ku = \pm\omega_1. \tag{4.6}$$

This equation describes beam space-charge waves (3.19): the plus sign corresponds to the fast wave ($\omega/k > u$) and the minus sign, to the slow wave ($\omega/k < u$). By virtue of (4.4), the energy density of these waves is

$$W = \pm\frac{\overline{E^2}}{4\pi} \frac{\omega}{\omega_1} \tag{4.7}$$

where the plus and minus signs refer to the fast and slow waves, respectively. We see that the fast wave carries the positive energy, and the slow wave carries the negative energy.

Both the magnitude and the sign of the wave energy depend on the choice of reference frame. For example, the total wave energy density in the laboratory reference frame for the Langmuir wave in a plasma at rest ($\omega = \omega_p$, $\varepsilon = 1 - \omega_p^2/\omega^2$) is

$$W = 2\frac{\overline{E^2}}{8\pi} = \frac{\overline{E^2}}{4\pi} \tag{4.8}$$

(the wave energy is split equally between the electric energy of the field and the mechanical energy of oscillation of particles). Absolutely the same result would be obtained for both space charge waves if we treated them in the beam's reference frame (where they are simply Langmuir beam oscillations). In the laboratory reference frame, however, in which the beam is in motion, the beam wave energy is different for the two waves and for one of them it is negative (although the electric field energy is equal, as before, to $\overline{E^2}/8\pi$). If both waves (of equal amplitude) are excited simultaneously in a quasineutral beam, their total energy is, as follows from (4.7),

$$W = \frac{\overline{E^2}}{4\pi\omega_1}(\omega_+ - \omega_-) = 2\frac{\overline{E^2}}{4\pi} \tag{4.9}$$

where ω_+ and ω_- are the frequencies of the slow and the fast waves, respectively.

To clarify the physical meaning of the results obtained above, let us look at a different derivation of expression (4.7), without resorting to definition (4.1) [4.3]. I have already mentioned that the wave energy density W is the sum of the energy density of the electric field, $W_E = \overline{E^2}/8\pi$, and the density of the kinetic energy W_K of particle oscillation. The latter is equal to the change in the beam kinetic energy density caused by the wave (averaged over one oscillation period),

$$W_K = \frac{m}{2}\overline{(u + v)^2(n_1 + n)} - \frac{m}{2}u^2 n_1 \tag{4.10}$$

where v and n are the perturbations of velocity and density of beam electrons due to the wave (in the longitudinal oscillations that we consider here, $v\|u\|E$); the overbar again denotes averaging over one oscillation period. The quantities v and n can be found using the equation of motion and Poisson's equation

$$\frac{d(u + v)}{dt} \equiv \frac{dv}{dt} \equiv \frac{\partial v}{\partial t} + u\frac{\partial v}{\partial z} = \frac{e}{m}E$$

and div $\boldsymbol{E} \equiv ikE = 4\pi ne$. Taking into account that the quantities v, n and E are harmonic functions of time, we obtain

$$\left.\begin{aligned} v &= i\frac{e}{m}\frac{E}{\omega - ku} \\ n &= \frac{1}{4\pi e}ikE \end{aligned}\right\} \tag{4.11}$$

or, in view of (4.6),

$$\frac{v}{n} = \pm \frac{4\pi e^2}{mk} \frac{1}{\omega_1} \qquad (4.12)$$

where the plus sign refers to the fast wave and the minus sign, to the slow one. We see that velocity and beam density perturbations in the slow wave are in antiphase; it will presently be shown that this behaviour determines the negative sign of kinetic energy density W_K and of the total energy density $W = W_K + W_E$ of this wave (in the fast wave, the perturbations v and n vary in phase and the quantities W_K and W are positive) [4.4]. According to (4.10), we find for the slow wave

$$W_K = \frac{m}{2} \left\{ \left[\frac{1}{T} \int_0^T (n_1 - n_0 \sin \omega t)(u + v_0 \sin \omega t)^2 \, dt \right] - n_1 u^2 \right\} \qquad (4.13)$$

where n_0 and v_0 are the amplitudes of the perturbations n and v (note that since the averaging of (4.13) is performed over one period T of oscillations, the result will definitely remain unchanged if the signs of addition and subtraction in parentheses in the integrand are reversed simultaneously). As follows from (4.13), (4.11) and (4.12), we have

$$W_K^S = -\frac{E_0^2}{8\pi} \frac{ku}{\omega_1} + \frac{E_0^2}{16\pi} \qquad (4.14)$$

where E_0 is the amplitude of the field E, which determines the amplitudes n_0 and v_0 according to (4.11), and the superscript 'S' refers to the slow wave. Hence, the total energy density of the slow wave is

$$W^S = W_K^S + \frac{E_0^2}{16\pi} = -\frac{E_0^2}{8\pi} \frac{\omega}{\omega_1} = -\frac{\overline{E^2}}{4\pi} \frac{\omega}{\omega_1} \qquad (4.15)$$

(where we have taken into account relation (4.6)). In calculating the kinetic energy density W_K^F of the fast wave, it is necessary, according to (4.12), to choose identical signs, in parentheses in the integrand of (4.13). This gives

$$W_K^F = \frac{E_0^2}{8\pi} \frac{ku}{\omega_1} + \frac{E_0^2}{16\pi} \qquad (4.16)$$

$$W^F = W_K^F + \frac{E_0^2}{16\pi} = \frac{\overline{E^2}}{4\pi} \frac{\omega}{\omega_1} \qquad (4.17)$$

We see that expressions (4.15)–(4.17) for both waves of the beam, obtained with simple kinematic arguments, coincide exactly with expression (4.7) whose derivation operated with the general definition (4.1) of the wave energy.

Now we can write the energy relations for the density of the total energy of the (beam+wave) system. The sum of the kinetic energy density of beam electrons T and the energy density $\overline{E^2}/4\pi$ of the electric field of the wave undergoes the following change caused by the excitation of the wave:

$$\left.\begin{array}{l} T^F + \dfrac{\overline{E^2}}{8\pi} = T^0 + W^F \\[3mm] T^S + \dfrac{\overline{E^2}}{8\pi} = T^0 - |W^S| \end{array}\right\} \qquad (4.18)$$

where the superscripts F and S specify the waves present and the superscript 0 indicates that no waves are excited. We see that if the fast wave appears, the total energy of the (beam+wave) system is greater than the original beam energy, and if the slow wave is excited, the total energy is lower than the original energy; the energy difference indicated by (4.18) is exactly equal to the absolute value of the energy of the excited waves. This is the physical meaning of the positive and negative wave energies. As for the energy of the electric field of oscillations (not the total oscillation energy with density W), it is certainly positive and increases with increasing wave amplitude: $W_E^S = W_E^F = E_0^2/16\pi$.

The derivation of the expression for W_K given above clarifies another factor of critical importance: the energy of one of the Langmuir waves in the medium is negative only because this medium (in this particular case, the beam) is moving. This is immediately seen from relations (4.14) and (4.16): if we set $u = 0$, the kinetic and the total energies of the slow wave are positive. For a plasma which is at rest as a whole, (4.14) and (4.16) imply

$$\left.\begin{array}{l} W_E = W_K = \overline{E^2}/8\pi \\[3mm] W = W_E + W_K = \overline{E^2}/4\pi \end{array}\right\} . \qquad (4.19)$$

This result signifies that the total energy of Langmuir oscillations of a plasma at rest is positive and splits into two equal parts: kinetic energy of oscillations of electrons and the energy of the electric field. The same result, (4.19), can be readily obtained by calculations using formulas (4.1) and (4.4).

These space-charge waves possess momentum [4.5, 4.6]. By analogy to definition (4.13) of the wave energy density, the wave momentum density is defined as the change in the momentum of beam electrons due to the wave (averaged over one oscillation period and referred to a unit beam volume):

$$P = \overline{m(n_1 + n)(u + v)} - mn_1 u = m\overline{nv}. \qquad (4.20)$$

By virtue of (4.11), (4.13) and (4.6), we have

$$P = \frac{W}{v_{\text{ph}}} = \frac{\pm(\omega/\omega_1)\overline{E^2}/4\pi}{v_{\text{ph}}} = \frac{\pm(\omega/\omega_1)\overline{E^2}/4\pi}{u \pm \omega_1/k} = \frac{n_1 m v^2}{v_{\text{ph}} - u} \qquad (4.21)$$

where $v_{\mathrm{ph}} = \omega/k$ is the phase velocity of the wave and the plus and minus signs refer to the fast and the slow waves, respectively. We see that momentum carried by the fast wave $(v_{\mathrm{ph}} > u)$ is positive, and that carried by the slow wave $(v_{\mathrm{ph}} < u)$ is negative. If both waves are simultaneously excited in the system (and have equal amplitudes), their total momentum is zero

$$P_\Sigma = W^{\mathrm{F}}/v_{\mathrm{ph}}^{\mathrm{F}} + W^{\mathrm{S}}/v_{\mathrm{ph}}^{\mathrm{S}} = 0 \qquad (4.22)$$

and their total energy is positive

$$W_\Sigma = W^{\mathrm{F}} + W^{\mathrm{S}} = \frac{\overline{E^2}}{4\pi\omega_1}(\omega^{\mathrm{F}} - \omega^{\mathrm{S}}) = 2\frac{\overline{E^2}}{4\pi}. \qquad (4.23)$$

This result corresponds to the fact that the system consisting of a slow and a fast beam space-charge wave is stable.

4.2   Instability in Interaction Between Waves with Opposite Signs of Energy

The derived formula (4.18) shows that additional energy must be introduced into the beam to excite in it the fast space-charge wave while energy must be extracted to excite the slow wave. In other words, the pumping of the slow wave requires a channel to drain its energy. Energy transfer to another wave—connected with the medium at rest and thus carrying positive energy (for example, to Langmuir oscillations of a plasma at rest or to an electromagnetic wave in a slowing-down HF structure)—can be such a channel.

Let us consider both these possibilities. If the electron beam is run through the plasma in such a way that the slow wave of the beam transfers its energy to the plasma wave, then both these waves are pumped (at the expense of the kinetic energy of the beam, of course), that is, the so-called hydrodynamic beam instability sets in. It will be shown below that the beam instability criteria found in chapter 3 can be easily obtained if one accepts the concept of negative-energy waves.

As a first example, we take one-dimensional Langmuir oscillations in the beam–plasma system [3.20]:

$$\varepsilon(\omega) = 1 - \frac{\omega_1^2}{(\omega - ku)^2} - \frac{\omega_{\mathrm{p}}^2}{\omega^2}. \qquad (4.24)$$

For a system to become unstable, it is necessary that the slow beam wave possesses a sufficiently large 'reservoir' from which to draw energy for pumping the plasma wave, that is, to compensate for the positive contribution of the plasma wave by the negative contribution of the slow wave to the total oscillation energy of system (4.1), that is, by virtue of (4.4),

$$-\frac{\omega \omega_1^2}{(\omega - ku)^3} = \frac{\omega_p^2}{\omega^2}$$

or

$$-\frac{(\omega - ku)}{\omega} = \left(\frac{n_1}{n_2}\right)^{1/3}. \tag{4.25}$$

This expression implies a beam instability threshold which coincides with that obtained in the 'conventional' analysis ($\varepsilon = 0$, $\partial \varepsilon / \partial \omega = 0$, see section 3.2). It is also important for further analysis that the frequency of oscillations close to the threshold, given by (3.33) as

$$\omega = \frac{ku}{1 + (n_1/n_2)^{1/3}}$$

satisfies the inequality

$$u > \omega/k \tag{4.26}$$

which is also obeyed by the slow wave of beam space charge; inequality (4.26) indicates that the beam 'leaves the excited wave behind'. (We discuss here the instability close to the threshold, where the increment is much lower than its maximum value (3.34)—see the concluding part of this section.)

Note also that it is no accident that relation (4.26) coincides with auto-phasing condition (3.38) which is the necessary condition of beam instability.

It is easy to show in a similar manner that both these approaches outlined above yield identical results for other beam instabilities (Buneman, beam-drift, etc): the beam instability mechanism is stipulated (in the hydrodynamic approximation we chose!) by the active coupling of the slow beam space-charge wave, which carries negative energy, to the wave in the medium, which carries positive energy.

In electronics, this active coupling of waves with opposite signs of energy is used for the generation and amplification of HF oscillations in systems with so-called slowing-down structures [4.7]. For example, the role of this structure in devices of the type of travelling-wave tubes (TWT) [1.6] is played by the metal helix surrounding the beam; the pitch of the helix is such that the axial velocity of the electromagnetic wave excited by the beam in the helix is very close to the beam velocity. To be precise: it must be equal to the phase velocity of the slow wave of the beam space charge. In this case, the generation (amplification) of oscillations is also caused by the active coupling of waves with opposite signs of energy: the wave in the fixed helix carries positive energy, that is, represents a dissipative load for the slow beam wave, and both waves grow as they propagate along the helix. This

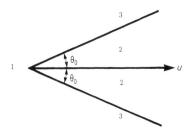

Figure 4.1 Spatially separated regions of normal and anomalous Doppler effect [4.9]. 1, region of normal Doppler effect; 2, region of anomalous Doppler effect; 3, Čerenkov cone.

takes place when (4.26) is satisfied: the beam overtakes the wave in the spiral. It is easy to see that the sum of energies of both enhanced waves remains constant in the process of propagation along the helix: the negative energy wave grows at the expense of energy loss to the wave excitation in the structure. This result, similar to relation (4.9), is known in electronics as the Chu theorem on kinetic power [4.7].

4.3   Elementary Processes at the Foundation of Beam Instabilities

Beam instabilities are based on three main elementary processes: the Vavilov–Čerenkov effect, the normal Doppler effect and the anomalous Doppler effect (ADE) [1.12]. The first two are quite familiar, so I will clarify here the physical meaning of the third process [4.8–4.10] (see figure 4.1). Let an oscillator possessing kinetic energy T of translational motion and internal energy U move along a direction z and emit an energy quantum $\hbar\omega \ll T$ at an angle Θ to this direction; the quantum has a momentum $\hbar k$. As a result of the 'recoil', the kinetic energy of the emitting oscillator is reduced by the quantity $\Delta T \simeq \hbar k_z u = \hbar k u \cos \Theta$, whence

$$\frac{\Delta T}{\hbar\omega} \simeq \frac{k_z u}{\omega} = \frac{u}{\omega/k_z} = \frac{u \cos \Theta}{v_{\text{ph}}} \qquad (4.27)$$

where $v_{\text{ph}} = \omega/k$ is the phase velocity of the wave in the medium. If $u > \omega/k_z$ (condition (4.26)), then

$$|\Delta T| > \hbar\omega \qquad (4.28)$$

that is, the system loses more energy than is removed by the quantum emitted (sic). This is one of the apparent paradoxes of the 'super-light-velocity' motion (in optics, this is the case of $u \cos \Theta > c/N$ where

$N = c/v_{\text{ph}}$ is the refractive index of the medium). To make this radiation possible, the excess of the energy lost must go to increase the internal energy of the system:

$$\Delta U = \hbar\omega\left(\frac{k_z u}{\omega} - 1\right) = \hbar\omega\left(\frac{u\cos\Theta}{v_{\text{ph}}} - 1\right). \qquad (4.29)$$

If conditions (4.26) and (4.28) are met, the system emits energy and thereby lifts itself to a higher, excited state (sic). This is the phenomenon known as the anomalous Doppler effect. It differs from the normal ('ordinary', 'sub-light-velocity') Doppler effect 'only' in the sign of ΔU in formula (4.29): the normal Doppler effect obeys inequalities that are inverse relative to (4.26) and (4.28); in the latter case, (4.29) implies that the emission of a quantum ('as usually') occurs at the expense of the internal energy of the system. The borderline between the normal and anomalous Doppler effects is the situation in which the Čerenkov condition holds:

$$u = \omega/k_z.$$

Also, $\Delta U = 0$: the emission of a quantum at the Čerenkov angle $\Theta_0 = \arccos(v_{\text{ph}}/u)$ does not change the internal energy of the system. As a result, the Vavilov–Čerenkov radiation can be emitted, for example, by a free charged particle having no internal degrees of freedom. All three effects are thus possible in the 'super-light-velocity' motion of the system $(u > \omega/k)$, depending on the direction of emission: if $\Theta > \Theta_0$—the normal Doppler effect; if $\Theta = \Theta_0$—the Vavilov–Čerenkov effect; if $\Theta < \Theta_0$—the anomalous Doppler effect. Only the normal Doppler effect is possible in the 'sub-light-velocity' motion $(u < v_{\text{ph}})$.

Let us consider a particular case in which the oscillator is a charged particle (e.g., an electron) moving freely at a velocity u along the external magnetic field $(H = H_z)$ and having a small transverse velocity component (perpendicular to H).

The rotation energy of this particle in the field H (the internal energy of the oscillator) changes by quanta:

$$\Delta U = n\hbar\omega_H \qquad \omega_H = eH/mc.$$

Here m is the particle mass, $n = 0, \pm 1, \pm 2, \ldots$ By virtue of (4.29), which expresses energy conservation,

$$\left.\begin{array}{l} n\hbar\omega_H = \hbar(\omega - k_z u) \\ (\omega - k_z u) = n\omega_H \end{array}\right\}. \qquad (4.30)$$

The case of $n > 0$ corresponds to the normal Doppler effect: the radiation frequency $(\omega - k_z u)$ in the case of the oscillator's reference frame equals the frequency of the corresponding quantum transition. The case of $n < 0$ corresponds to the anomalous Doppler effect, $\hbar k_z u = \hbar \omega + \hbar |n| \omega_H$: the change in the kinetic energy of the longitudinal motion of the oscillator is spent on the emission of a quantum $\hbar \omega$ and on an increase in the internal energy of the oscillator, that is, it increases the energy of its rotation in the magnetic field. If the particle has no rotational energy before emission, it 'spins' as a result of the emission of a quantum (if $n < 0$). For a particle beam, the anomalous Doppler effect facilitates the isotropization of the beam, in agreement with the experimental observations [4.11, 4.12]. Finally, the case of $n = 0$ corresponds to the Vavilov–Čerenkov effect; the emission of a quantum is not accompanied by a change in the rotational energy of the particle.

4.4   Analogy of the Induced Anomalous Doppler Effect to the Instability of a Negative-Energy Wave

It can be shown that there is a profound analogy of the anomalous Doppler effect discussed above to the mechanism of active coupling of waves with opposite signs of energy [4.3]. Indeed, when a negative-energy wave pumps ('emits') a positive-energy wave, the former wave transforms to a state of higher oscillation amplitude, that is, of higher electric field. On the other hand, a system moving faster than the wave that it is emitting, is transformed by the anomalous Doppler effect to a higher excitation state, having higher internal energy. As far as dispersion is concerned, namely, as long as condition (4.26) is satisfied ($u > \omega/k$), both these phenomena take place in the same situation. Thus the dispersion of the slow wave of the beam space charge, which carries the negative energy and undergoes instability in a medium with a positive energy wave, is described by equation (4.6),

$$\omega - k_z u = -\omega_1$$

which is completely analogous to dispersion (4.30) in the case of the anomalous Doppler effect:

$$\omega - k_z u = -|n| \omega_H.$$

These facts can be supplemented with another example: the so-called electron beam cyclotron waves which are widely used in microwave electronics and plasma physics. If beam electrons have only a velocity component along the magnetic field ($u = u_z$), the dispersion of these (electrostatic) waves is determined by the dielectric constant

$$\varepsilon = 1 - \frac{\omega_1^2}{(\omega - k_z u)^2 - \omega_H^2} \qquad (4.31)$$

whence we find the natural wave frequencies $(\varepsilon = 0)$

$$\left.\begin{array}{l} \omega - k_z u = \pm(\omega_H^2 + \omega_1^2)^{1/2} \\ \omega - k_z u \simeq \omega_H \end{array}\right\} \qquad (4.32)$$

where the plus sign corresponds to the fast wave and the minus sign, to the slow wave. We see that the dispersion of cyclotron waves practically coincide with condition (4.30) of emission by the Larmor oscillator. We find from (4.31) and (4.1) that the fast wave corresponding to the normal Doppler effect $(u < \omega/k_z)$ carries positive energy, while the slow wave corresponding to the anomalous Doppler effect $(u > \omega/k_z)$ carries negative energy. If $\omega_1 \gg \omega_H$, we arrive at the space charge waves described in (4.6):

$$\omega - k_z u = \pm\omega_1.$$

The ensemble of facts as presented here demonstrates convincingly that there is a physical analogy between the beam instability due to the active coupling of waves with opposite signs of energy, on one hand, and the anomalous Doppler effect, on the other. This effect (this time the induced effect, not the spontaneous one, of course) lies at the foundation of all beam instabilities discussed in chapter 3. Also, this is the effect which is realized in the hydrodynamic regime of beam instability when the beam auto-phasing condition (3.38) is satisfied. (For details of the problems discussed, see also [3.10, 3.18, 4.13].

4.5 Effect of the Spread of Electron Beam Velocities on the Type of Beam Instability

We apply the hydrodynamic approximation throughout the book to analyse the physics of collective beam–plasma interaction. A rigorous kinetic analysis [3.20] shows that this approximation is quite stringent for beams with a sufficiently small thermal energy spread (the smallness criterion is given in this section).

Let us dwell now on the aspect of principal importance: the effect of velocity spread on the characteristics of beam instability. The anomalous Doppler effect occurs in the 'hydrodynamic' beam regime, with sufficiently small velocity spread, in which all beam electrons move faster than the wave they pump $(u > \omega/k_z)$, that is,

$$\Delta u < (u_0 - \omega/k_z) \qquad (4.33)$$

(see figure 4.2). The following question arises: what if the beam velocity spread ceases to be 'sufficiently small' and condition (4.33) does not

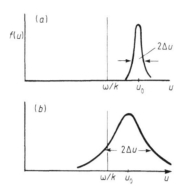

Figure 4.2   (*a*) Hydrodynamic beam regime: $\Delta u < (u_0 - \omega/k_z)$; (*b*) kinetic beam regime: $\Delta u > (u_0 - \omega/k_z)$. The plot assumes $k_z = k$.

hold? We will answer this question for the case of one-dimensional motion ($k = k_z$). It is found that in the regime of large velocity spread (this regime is known as the kinetic regime) the situation changes drastically: first, the anomalous Doppler effect vanishes, and second, the negative-energy beam wave simultaneously ceases to be excited, that is, it damps out. This coincidence, that is, simultaneous vanishing of the anomalous Doppler effect and of the instability (the pumping) of the negative-energy wave is definitely not accidental; it is an indication of a profound physical analogy existing between these two phenomena.

Let us define the boundary between the hydrodynamic and the kinetic regimes of beam instability; to be specific, we choose one-dimensional geometry. The boundary corresponds to such maximal velocity spread Δu_{max} of beam electrons at which the contribution of the slow wave of the beam space charge to the total energy of the beam–plasma wave system is still capable of compensating for the contribution of the wave in the medium. According to this definition we find, similarly to (4.25),

$$\frac{\Delta u_{max}}{u_0} \simeq \frac{\omega_c}{ku_0} \left(\frac{\omega_1^2}{\omega_2^2}\right)^{1/3} \tag{4.34}$$

where ω_c is the oscillation frequency close to the threshold and ω_2 is the natural (Langmuir) oscillation frequency of the fixed (as a whole) component of the system. (A similar criterion can be derived for the cyclotron wave.) For example, we have for the electron–electron instability (section 3.3) that $\omega_2 = \omega_p$ and

$$\frac{\Delta u_{max}}{u_0} \simeq (n_1/n_2)^{1/3}. \tag{4.35}$$

For the electron–ion instability (section 3.2),

$$\frac{\Delta u_{\max}}{u_0} \lesssim 1. \tag{4.36}$$

In fact, these results are directly implied by (4.33) and (3.12).

I will, in addition, comment on the physical meaning of the criterion of smallness of the thermal spread in beam velocities, using as an example the hydrodynamic regime of 'ordinary beam' (electron–electron) instability. For the motion of beam particles to be coherent, it is necessary for a beam particle to shift during the time of one inverse instability increment $(1/\gamma)$ by a distance considerably smaller than one half wavelength (π/k):

$$\Delta u/\gamma \ll \pi/k. \tag{4.37}$$

From (4.37) together with (3.34), we obtain (4.35). If this criterion is violated, the motion of beam particles is not coherent, so that it is not the fields radiated by individual particles but their intensities that add up in the beam instability—as in non-coherent scattering (I will return to this aspect in chapter 11). This regime of (essentially weaker) beam instability is known as kinetic. It stems from the induced Vavilov–Čerenkov effect (or inverse Landau damping) and takes place if the Čerenkov condition is satisfied: $v_{\text{ph}} = u_0$.

One clarification is needed before we conclude this section. The matter of the sign of wave energy in the beam–plasma interaction (rather, in the excitation of the Langmuir wave by the electron beam) generated, some time ago, a lively discussion which was mostly caused by incorrect interpretation of the term 'negative-energy wave'. In fact, expression (4.1) for the wave energy can be used correctly only if the oscillation frequency ω is sufficiently well defined, that is, if the oscillation increment γ is sufficiently small. Since the uncertainty in ω approximately equals γ, the beam–plasma wave will have negative energy only if criterion (4.33) is satisfied in the replacement $\omega \rightarrow (\omega + \gamma)$, that is, when $k_z u - (\omega + \gamma) > \Delta u$. To have this, it is necessary that at least

$$\gamma < (ku - \omega_{\text{p}}) = \omega_{\text{p}} (n_1/n_2)^{1/3}$$

(see (4.25) and (4.26)). On the other hand, the maximal increment of beam instability is (provided the overcriticality is sufficient) $\gamma_{\max} \simeq \omega_{\text{p}}(n_1/n_2)^{1/3}$. This means that the energy of the beam–plasma wave under the conditions of maximum increment is positive[†]. This is natural since in

---

[†] These arguments, including the above expression for the instability increment, hold for a *tenuous* non-relativistic beam in a *dense* plasma. If, however, the beam is relativistic and at the same time sufficiently dense ($\gamma_0 n_1/n_2 \gg 1$), the beam instability is based on the anomalous Doppler effect (therefore, the wave energy is negative, as we have demonstrated in this section). This was shown in [4.14]; a similar result was reported in [3.10].

this case the beam propagates in the kinetic regime. The wave energy sign is negative only in the hydrodynamic-beam regime, in which $\gamma < \gamma_{max}$, for example, is close to the instability threshold (this has been explained above). In view of all these arguments, we can assume that the apparent discrepancies in the conclusions of a number of authors about the wave energy sign in the beam–plasma interaction have been explained away (see also [3.18]).

4.6   Effects of Dissipation and Collisions on Beam Instability in Plasmas

Dissipative processes cause damping of positive-energy waves. However, they may pump waves carrying negative energy. I will demonstrate this using as an example the so-called resistive microwave amplifier [4.15, 4.16]. In such amplifiers, the beam is surrounded with walls of finite (relatively low) conductivity, in order to produce a dissipative load. The electric field of the slow wave of the beam induces a conduction current in the walls, that is, Joule losses, which lead to pumping of the wave (obviously, at the expense of the kinetic energy of beam electrons). We can write the dispersion equation of this system. First, we find the current densities induced by the wave field in the beam (j_b) and in the walls (j_w). By virtue of (4.11),

$$j_b = n_1 e v = i n_1 e^2 E / m(\omega - ku)$$

and $j_w = \sigma E$, where σ is the conductivity. According to the continuity equation, the beam and wall space charge densities are given by the relations

$$-i(\omega - ku)\rho_b \equiv \frac{\partial \rho_b}{\partial t} = -\operatorname{div} j_b$$

$$-i\omega \rho_w \equiv \frac{\partial \rho_w}{\partial t} = -\operatorname{div} j_w.$$

Substituting the total space charge density $\rho = \rho_b + \rho_w$ into Poisson's equation $\operatorname{div} E = 4\pi\rho$, we obtain the dispersion equation

$$\frac{\omega_1^2}{(\omega - ku)^2} + 4\pi\sigma/i\omega = 1 \qquad (4.38)$$

whence

$$\omega - ku = \pm\omega_1(1 - i2\pi\sigma/\omega) \qquad (4.39)$$

where the plus sign in the right-hand side refers to the fast wave and the minus sign, to the slow wave. We see that the imaginary part of the slow wave is positive, which signifies the pumping of oscillations (an instability). In this example, as in all those discussed above, the instability arises if condition (4.26) is met: all beam electrons move faster than the wave they excite, that is, the conditions are those of the anomalous Doppler effect.

Let us now consider the effect of collisions between particles on the characteristics of beam instability. It may seem at first glance that since collisions of plasma electrons (among themselves or with other particles) perturb the ordered motion of particles in the plasma wave, they will lead to only one result: the damping of plasma oscillations. However, this conclusion is generally incorrect.

In order to understand the role of collisions in the beam instability dynamics, let us return to the derivation of the dispersion equation of electron oscillations (see chapter 3); for simplicity, we consider the one-dimensional case. In this derivation, we need to take into account only one correction: since the collisions of plasma electrons cancel the oriented momentum in a time of the order of $\tau = 1/\nu$ (where ν is the collision frequency), they produce the friction force $-mv/\tau = -mv\nu$. Taking this force into account, we transform the equations of motion of a plasma electron to the form

$$m\frac{\mathrm{d}v_e}{\mathrm{d}t} = -eE_0\exp[\mathrm{i}(kz - \omega t)] - mv\nu$$

or

$$-\mathrm{i}\omega mv = -eE - mv\nu$$

whence $v_e = eE/\mathrm{i}m(\omega + \mathrm{i}\nu)$ and the current oscillation density is $j_e = en_ev_e = n_0e^2E/\mathrm{mi}(\omega + \mathrm{i}\nu)$. The oscillatory electron charge density given by the continuity equation $\partial\rho_e/\partial t = -\operatorname{div} j_e$ then equals $\rho_e = n_ee^2\operatorname{div}E/m\omega(\omega + \mathrm{i}\nu)$ (this expression then transforms into the familiar one, $\rho_e = n_ee^2\operatorname{div}E/m\omega^2$, as $\nu/\omega \to 0$). Hence, the dispersion equation of free oscillations of plasma electrons (without the external drive, that is, in the absence of the particle beam) takes the form

$$\omega_\mathrm{p}^2/\omega(\omega + \mathrm{i}\nu) = 1 \qquad (4.40)$$

whence

$$\omega = \omega_\mathrm{p} - \mathrm{i}\nu/2 \qquad (4.41)$$

where we assumed that $\nu \ll \omega_\mathrm{p}$. We see that if there is no source of oscillation pumping, oscillations damp out: their amplitude decreases with time with decrement $\nu/2$ and the intensity (amplitude squared) damps out with decrement ν. This is a natural result: the plasma wave carries positive energy and therefore dissipation causes the wave to damp out. However, if a monoenergetic beam is passed through the plasma, and a slow space-charge wave carrying negative energy is excited in the beam, then collisions of plasma particles (the dissipation of beam energy) pump up this wave, this is similar to processes in the dissipative microwave amplifier.

As a result, the plasma wave is enhanced. This phenomenon is known as the dissipative instability; it survives as long as the beam regime is hydrodynamic and the beam satisfies the small velocity spread condition (4.33). If the velocity spread exceeds threshold (4.33) (the kinetic regime), the energy of (both) space charge waves is positive and collisions ultimately damp out the plasma oscillations. This fact can be rephrased as follows: if particle collisions in the wave pump the wave, the wave energy is negative, and if the wave is damped out, the wave energy is positive.

4.7   On Negative-Energy Waves in Hydrodynamics

Negative-energy waves were also described in hydrodynamics [4.17–4.20]. This happened about one year earlier than in plasma physics [4.2], although the concept itself was then forgotten in hydrodynamics for 15 years; moreover, an (erroneous) opinion was ventured that this phenomenon was impossible (see [4.19, 4.20]).

Hydrodynamics also has instabilities that can be interpreted as resulting from an active coupling of negative and positive energy waves. One of the most impressive examples is the classical Kelvin–Helmholtz instability (KH), that is, the instability of tangential 'discontinuity' of velocity. It arises precisely when the branch of positive-energy waves intersects the negative-energy wave branch, and the positive-energy wave serves as a 'dissipative load' for the negative-energy wave [4.18–4.20]. By analogy with the discussion in section 4.4, the instability—regarded as a consequence of the active coupling of waves with opposite signs of energy—can be interpreted as the collective (driven) anomalous Doppler effect: the negative-energy wave 'emits' a positive-energy wave and, in response to this emission, is enhanced [4.3].

The energy source is, in this case, the flow of the fluid with a velocity jump. An interesting example of anomalous Doppler effect (instability due to the intersection of negative-energy and positive-energy branches) is found in a familiar aerodynamic effect: the flutter, which is related to a specific type of oscillation of airfoil. It has been discussed in the framework of the anomalous Doppler effect in [4.10]; it has also been analysed in detail in terms of an analogy of this effect to the instability of negative-energy waves [4.21].

We thus find that in the region under consideration, there is a profound analogy of phenomena in plasma physics and in hydrodynamics.

Another impressive example of this analogy is connected with the phenomenon known as 'Landau damping' (see, e.g., [3.24, 4.22]). This is the damping of plasma waves due to their Čerenkov absorption by 'resonant' particles moving at velocities equal to the phase velocity of the (absorbed) wave. If the number of particles moving 'slightly' slower than the wave is greater than the number of particles moving 'slightly' faster than the wave, Landau damping takes place, and if these inequalities between the slower and the faster particles are reversed, the wave is kinetically pumped via the Vavilov–Čerenkov effect. In fact, it has been shown quite recently (see [3.24, 4.22]) that a classical hydrodynamic phenomenon such as the instability of flow with non-uniform velocity profile containing an inflection point (the Rayleigh criterion [3.24]) is physically similar to the Čerenkov pumping of waves by resonant particles (i.e., to 'inverse Landau damping'). At the inflection point, we find an exact resonance of the particles and the wave. The presence of the inflection point (where the second derivative undergoes sign reversal) signifies that the numbers of particles in resonance with the wave are not equal on the two sides of this point. If the flow vorticity at the inflection point goes through a maximum, the number of resonant particles which 'slightly' overtake the wave is greater than the number of those which 'slightly' lag behind the wave, so that resonant particles pump the wave. In this case, the Rayleigh criterion corresponds to instability.

The instability of a non-uniform-density hollow-rod beam with non-compensated space charge, propagating along the magnetic field, has a similar physical meaning. The space charge of the beam creates an electric field perpendicular to the external magnetic field. The crossed fields E and H drive the rotation of the beam. In the present case, this rotation has a non-uniform velocity profile which is unstable under certain conditions (see section 3.6). This instability is known as diocotron instability. Examples of experimental observation of diocotron instability are given in section 6.10.

It is of special interest that some flows that have no inflection point on the velocity profile, also manifest instability under certain conditions. An example is plane Poiseuille flow. As Heisenberg established in one of his first papers, this instability is dissipative: it is stimulated by viscosity. In view of our current knowledge, it can be stated that this effect is caused by negative-energy waves [3.24].

For additional information on the analogy of the relevant phenomena in hydrodynamics and plasma physics, see [3.25–3.29].

References

[4.1]   Landau L D, Lifshitz E M and Pitaevsky L P 1984 *Electrodynamics of Continuous Media* 2nd edition (Oxford: Pergamon)

[4.2]   Kadomtsev B B, Mikhailovsky and Timofeev A V 1964 *Zh. Eksp. Theor. Fiz.* **47** 2266 (Engl. Transl. 1965 *Sov. Phys.–JETP* **20** 1517)

[4.3]   Nezlin M V 1976 *Uspekhi Fiz. Nauk* **120** 481 (Engl. Transl. 1976 *Sov. Phys.–USPEKHI* **19** 946)

[4.4]   Briggs R J 1971 *Two-Beam Instability* In *Advances in Plasma Physics* vols 3 ed A Simon and W B Thompson (New York: Wiley Interscience)

[4.5]   Pierce J R 1961 *J. Appl. Phys.* **32** 2580

[4.6]   Pierce J R 1974 *Almost All About Waves* (Cambridge, Mass.: MIT)

[4.7]   Louisell W H 1960 *Coupled Mode and Parametric Electronics* (New York: Wiley)

[4.8]   Ginzburg V L and Frank I M 1947 *DAN SSSR* **56** 583, 699

[4.9]   Ginzburg V L 1989 *Applications of Electrodynamics and Theoretical Physics and Astrophysics* (New York: Gordon and Breach)

[4.10] Tamm I E 1959 *Uspekhi Fiz. Nauk* **68** 387

[4.11] Shustin E G, Popovich V P and Kharchenko I F 1970 *Zh. Eksp. Theor. Fiz.* **59** 657 (Engl. Transl. 1971 *Sov. Phys.-JETP* **32** 358)

[4.12] Gorozhanin D V and Ivanov B I 1987 *Preprint FTI AN USSR no 87-6* (Moscow: TSNIIAtominform)

[4.13] Dendy R O, Lashmore-Davies C N and Montes A 1986 *Phys. Fluids* **29** 4040

[4.14] Bliokh Yu P, Karas V I, Lyubarsky M G, Onishchenko I N and Fainberg Ya B 1984 *DAN SSSR* **275** 56 (Engl. Transl. 1984 *Sov. Phys.-Doklady* **29** 202)

[4.15] Birdsall C K, Brewer G R and Haeff A V 1953 *Proc. IRE* **41** 865

[4.16] Lopukhin V M and Vedenov A A 1954 *Uspekhi Fiz. Nauk* **53** 69

[4.17] Benjamin T B 1963 *J. Fluid Mech.* **16** 436

[4.18] Cairns R A 1979 *J. Fluid Mech.* **92** 1

[4.19] Ostrovsky L A, Rybak S A and Tsimring L Sh 1986 *Uspekhi Fiz. Nauk* **150** 417

[4.20] Stepanyants Yu A and Fabrikant A L 1989 *Uspekhi Fiz. Nauk* **159** 83

[4.21] Nemtsov B E 1985 *Izv. VUZov, Radiofizika* **28** 1549

[4.22] Andronov A A and Fabrikant A L 1979 *Nonlinear Waves* ed A V Gaponov-Grekhov (Moscow: Nauka) p 68 (in Russian)

5 Electron, Ion and Plasma Beams in Laboratory Plasma

5.1 Neutralization of Electron Beam Space Charge, Quasineutral Beams, Plasma Beams

An intense charged particle beam can freely propagate through a medium if its space charge is neutralized by charges of the opposite sign. If the beam propagates through a neutral medium, this neutralization is a result of gas ionization by the beam. The neutralizing particles then accumulate in the beam while the oppositely charged particles are repelled from the beam by its electric field. Under certain conditions, the electric field of the beam may almost vanish.

Let a monoenergetic electron beam of density n_1 propagate at a velocity u through a neutral gas with particle density n_0. The beam is shaped into a cylindrical rod of length L and radius a and propagates along a strong magnetic field in the direction of the axis of an equipotential volume bounded by metallic walls: we assume the Larmor radius of ions to be small in comparison with the beam radius (see figure 2.1). In the stationary state, the beam has an equilibrium potential φ with respect to the walls; its absolute value and sign are implied by the relation between the rate of ion generation in the beam, $n_1 n_0 \sigma_i u \pi a^2 L$, and the free ion flux along the magnetic field, $2 n_+ v_+ \pi a^2/4$, where σ_i is the effective cross-section of ionization of the gas by the beam electrons; n_+ and v_+ are the ion density and the mean ion velocity, respectively. If the ion production is less intense than the free ion flux, the beam potential φ is negative and ion concentration builds up in the beam until the state of quasineutrality sets in: $n_+ \simeq n_1$. In this state, the beam consists almost totally of two components, namely, the fast (primary) electrons and the neutralizing ions; the third component, that is, the slow (secondary) electrons, is virtually non-existent. We refer to this two-component system of charged particles as a *quasineutral electron beam*. Hence, the condition that the beam potential is negative and the beam is quasineutral can be written as

$$n_0 \sigma_i u L < \frac{v_+}{2}$$

or

$$1/n_0 \sigma_i u > 2L/v_+ \tag{5.1}$$

which implies

$$\tau_i > \tau_+. \tag{5.2}$$

Here $\tau_i = 1/n_0 \sigma_i u$ is the so-called ionization time, that is, the time during which the number of ions that the beam creates in one unit volume is equal to the beam density. Indeed, if the ion production rate is $n_1 n_0 \sigma_i u$ ions per second, the ionization number over an arbitrary time t is $n_1 n_0 \sigma_i u t$, so that $n_1 n_0 \sigma_i u \tau_i = n_1$, whence we obtain the expression for τ_i given above. The quantity $\tau_+ = 2L/v_+$ is the average lifetime of an ion in a beam in free flight; it is implied by the balance equation,

$$n_+ V/\tau_+ = 2n_+ v_+ S/4$$

where $v = SL$ is the beam volume and S is its cross-sectional area. Let ions and neutral atoms be hydrogen, $L = 100\,\text{cm}$, $v_+ = 2.5 \times 10^6\,\text{cm/s}$ (the average ion energy is 3 eV), $u = 10^9\,\text{cm/s}$, $\sigma_i \simeq 10^{-16}\,\text{cm}^2$. Condition (5.2) is then satisfied if $n_0 < 1.25 \times 10^{11}\,\text{cm}^{-3}$, that is, the gas pressure must be

$$p < 2.5 \times 10^{-6}\,\text{mm Hg†.} \tag{5.3}$$

Furthermore, it is important that we take into account that the area through which ions leave the system is usually several times greater than the cross section πa^2 of the electron beam. Hence, if condition (5.3) is satisfied, the potential is negative (and the system is two-component) even in heavy gases (e.g., nitrogen and argon) with their large effective ionization cross section σ_i.

If condition (5.2) is not met, the equilibrium ion density is greater than the beam electron density: $n_+ > n_1$. The beam potential φ reverses its sign to positive, slow electrons are not lost from the beam, and the system of charged particles becomes a three-component system. If the gas pressure is sufficiently high, the density of the 'excess' plasma becomes much greater than that of the primary electron flux. By definition, we refer to a three-component system with a potential $\varphi > 0$ as a *plasma beam*. The borderline between the quasineutral and the plasma beam modes is the regime of $\varphi = 0$, when $\tau_i = \tau_+$.

---

† $1\,\text{mm Hg} = 1.33 \times 10^2$ Pa.

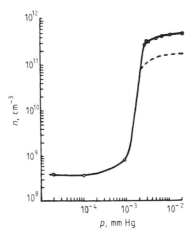

Figure 5.1   Plasma density as a function of gas pressure (H_2).
Beam–plasma discharge appears at $p \simeq 2 \times 10^{-3}$ mm Hg.

5.2   Beam–Plasma Discharge

The role of secondary electrons in gas ionization may become predominant
in a plasma beam in which the secondary electron density is greater than
that of primary electrons, and whose temperature is close to the gas
ionization potential or even higher than that. In this situation, the so-
called *beam–plasma discharge* develops [5.1–5.3]; while it sets in, the plasma
density may grow by several orders of magnitude as compared with that
produced by the ionization of the gas by primary beam electrons only.
Under the conditions of figure 5.1 (see [5.2]), the beam–plasma discharge
arises when the gas pressure exceeds a certain threshold ($p \simeq 2 \times 10^{-3}$ mm
Hg). The theory of this phenomenon is given in [5.3]. If plasma electrons
have a Maxwellian energy distribution, it is not difficult to ascertain
that the quantity $\langle \sigma_i v_e \rangle$, averaged over the energy distribution of plasma
particles, is

$$\langle \sigma_i v_e \rangle \simeq \sigma_m \bar{v}_e \exp\left(-J/T_e\right) \qquad (5.4)$$

where σ_m is the maximum cross section of gas ionization, J is the ionization
energy, $\bar{v}_e$ is the mean electron velocity, and $\overline{m v_e^2}/2 = 8 T_e/\pi$; this quantity
dictates the ionizing capacity of plasma electrons. Consequently, the ratio
of the rates of gas ionization by the plasma electrons and by the beam
electrons is

$$\frac{n_e \langle \sigma_i v_e \rangle}{n_1 \sigma_i u} = \frac{n_e}{n_1} \frac{\sigma_m}{\sigma_i} \frac{\bar{v}_e}{u} \exp\left(-J/T_e\right) \qquad (5.5)$$

where σ_i is the ionization cross section of the gas by beam electrons
whose energy is $mu^2/2$. For instance, if $\sigma_m/\sigma_i \simeq 5$, $n_e/n_1 \simeq 3 \times 10^2$,

$T_e = 2J$, $v_e/u \simeq 10^{-1}$, then (5.5) implies that the rate of ion production by secondary (plasma) electrons is higher by two orders of magnitude than by primary-beam electrons. In this case, the primary electrons act as 'seeds' of discharge. According to (5.5), the ionizing capacity of plasma electrons increases steeply as their concentration and temperature T_e are raised by the beam; hence, specific conditions can produce avalanche growth of plasma density. An increase in plasma density continues until this growth starts to reduce T_e, after which the plasma density reaches saturation.

In the plasma–beam discharge, plasma electrons are heated up via efficient pumping of Langmuir oscillations by the beam. However, it is possible to produce conditions under which the frequency of collisions of plasma electrons exceeds the increment of beam instability. The pumping of Langmuir waves is then suppressed while plasma electrons become mostly cooled in collisions with heavy particles. In this way, it is possible to generate the so-called *supercooled plasma*, possessing very unusual thermodynamic properties [5.4, 5.5]. In contrast to the plasma–beam discharge, the gas ionization by plasma electrons is negligible in this medium in comparison with the ionization by beam electrons. As the beam passes through such a medium, no collective phenomena are generated and the beam relaxation length is completely determined by elementary processes, that is, by the ionization losses of beam particles. Supercooled plasma is, more than anything, a borderline of the beam physics, since beam physics is associated nowadays mostly with collective processes [1.12–1.16]. (For more details on the physics of the beam–plasma discharge, see [5.6–5.12, 3.19] and references cited therein.)

5.3   On Temperature of Particles in Beam Plasma

The calculation of the energies of secondary particles in a plasma beam and a quasineutral beam is of principal importance. As for ions, their temperature T_i is determined not by those infinitesimal energies with which they are generated but by the potential difference across which ions pass when travelling in the beam. These potential differences in the plasma beam may be of the order of the electron temperature T_e, so that it is reasonable to assume for the evaluation that $T_i \lesssim T_e$. The electron temperature can be found from the balance equation

$$Q = NW_2 \tag{5.6}$$

where Q is the energy that the beam transfers to plasma particles (i.e. electrons) in one second, N is the flux of particles from the plasma to the walls, and W_2 is the mean energy of the particles leaving the plasma. Let us assume that over one relaxation length of the beam, it transfers to the plasma (via plasma waves) about one half of its energy: the theory says [5.6, 5.7] that the fraction of energy transferred falls between 1/3 and 3/4 (depending on the geometry of the system), with the energy transfer distributed equally between the mechanical (thermal) energy of plasma electrons and the electrostatic energy of oscillations. If the length of the plasma column is greater than the beam relaxation length, we can assume that about one fourth of the entire beam energy flux is spent on heating the plasma particles, that is, $1/4 n_1 (mu^2/2) u$ per $1\,cm^2$ of beam cross section. The flux of particles from the plasma is a combination of electron and ion fluxes. The density of the former is $\frac{1}{2} n_e v_e \exp(-\varphi/T_e)$, where φ is the equilibrium (positive) plasma potential with respect to the surrounding walls, necessary to sustain its quasineutrality. Connected with this potential is the electric field of the space-charge layer separating the plasma from the walls. This field penetrates the plasma and accelerates ions; by virtue of Bohm's theorem [5.13], ions arriving at the boundary of this layer have the energy of directed motion equal to one half of the electron temperature: $Mv_+^2/2 = T_e/2$, that is, their velocities are $v_+ = (T_e/M)^{1/2}$. The plasma density N_+ at this location is, by virtue of Boltzmann's law, less than the unperturbed plasma density by a factor of $e^{1/2} = 2.7^{1/2}$. (It is assumed that the ion temperature $T_+ \ll T_e$, as is usually the case.) The flux density of ions leaking from the plasma is $n_+ v_+$; electrons leave the plasma column through only one surface, which is opposite the electron gun, but ions escape from the plasma across both surfaces (for the time being, the ion flux across the lateral surfaces of the column is ignored). The electron and ion fluxes must be equal to maintain the quasineutrality of the plasma. Therefore, since $v_e = ((8/\pi)T_e/m)^{1/2}$, we obtain

$$\frac{n_e v_e}{4} \exp\left(-\frac{e\varphi}{T_e}\right) = 2 n_+ v_+ \quad \text{or} \quad \frac{e\varphi}{T_e} = \ln\left(\frac{2.7}{8\pi}\frac{M}{m}\right) \simeq \ln\left(\frac{M}{8m}\right)$$

The mean energy carried by a particle leaving the plasma is $e\varphi$, and the total energy flux lost by the plasma per unit surface is

$$Q \simeq 2 \cdot 2 n_+ v_+ e\varphi \simeq \left(\frac{2m}{M}\right)^{1/2} n_e v_e e\varphi \tag{5.7}$$

or

$$Q \simeq \frac{1}{2}\left(\frac{8m}{M}\right)^{1/2} \ln\left(\frac{M}{8m}\right)^{1/2} n_e v_e T_e. \tag{5.8}$$

Now the balance equation (5.6) changes to

$$\frac{1}{4} n_1 \frac{mu^2}{2} u \simeq \frac{1}{4}\left(\frac{8m}{M}\right)^{1/2} \ln\left(\frac{M}{8m}\right)^{1/2} n_e v_e T_e \tag{5.9}$$

whence

$$\frac{T_e}{W_1} \simeq \left[\frac{1}{12}\frac{M/m}{[\ln(M/m)]^2}\frac{n_1^2}{n_e^2}\right]^{1/3} \tag{5.10}$$

where $W_1 = mu^2/2$ is the beam electron energy. If $n_e = 10^2 n_1$, $M/m = 2 \times 10^3$ (hydrogen) and $W_1 = 150\,\text{eV}$, equation (5.10) gives $T_e \simeq 10\,\text{eV}$. If the ratio of the plasma density to the beam density is increased by an order of magnitude, we obtain $T_e \simeq 2\,\text{eV}$. The calculation above does not pretend to give high accuracy. In a more detailed calculation, one would have to take into account that as φ increases (i.e., as T_e increases), the effective surface of ion escape (ions are much less magnetized then electrons are) also grows considerably; the anomalous diffusion of electrons at right angles to the magnetic field† and some other factors would have to be included as well. As a result, we can point to an actual range of variation of T_e: from several eV to several tens of eV.

5.4   Hot-Cathode Discharge as a Means of Producing Plasma Beams

The hot-cathode gas discharge was used in the experiments described below for generating a plasma beam of sufficiently high density in a strong longitudinal magnetic field. Let us briefly review the main properties of such a discharge and the method of producing it [5.13–5.17] (figures 5.2, 5.3). The source of the electron beam is a hot cathode (e.g., a tungsten cathode heated by electron bombardment, or a hexaboride lanthanum cathode heated by the thermal radiation of a tungsten helix). The plasma is produced in a discharge chamber into which the working gas is fed; an accelerating potential difference is applied to electrons between the cathode and the chamber (the discharge voltage V_d). If the gas pressure in the discharge chamber exceeds a certain threshold, the discharge is ignited, that is, a dense beam of primary electrons and a dense plasma are formed, propagating along the magnetic field; in other words, a high-density plasma beam is generated. When the discharge is ignited, practically the entire discharge voltage is concentrated in a thin layer at the cathode. When the cathode temperature is sufficiently high, the space charge of electrons causes the electric field close to the cathode surface to vanish, and the current density of accelerated electrons obeys the well known Child-Langmuir 'law of three halves'

$$j_1 \simeq \frac{2\sqrt{2}}{9\pi} \left(\frac{e}{m} \right)^{1/2} \frac{V_d^{3/2}}{d^2} \tag{5.11}$$

where d is the cathode layer thickness. Since the accelerating field does not penetrate the plasma, the ion current from the plasma to the cathode is

† Anomalous plasma diffusion is the main problem of high-temperature plasma physics; it was treated as a special subject in a number of monographs (see, e.g., [1.15, 1.16].

given by a completely analogous relation:

$$j_+ \simeq \frac{2\sqrt{2}}{9\pi}\left(\frac{e}{M}\right)^{1/2}\frac{V_d^{3/2}}{d^2}. \tag{5.12}$$

As a result, the hot-cathode discharge at a sufficiently high temperature is described by the Langmuir formula

$$\frac{j_+}{j_1} \simeq \left(\frac{m}{M}\right)^{1/2}. \tag{5.13}$$

If the cathode temperature is insufficiently high, the current density is lower than the quantity given by (5.11), so that

$$\frac{j_+}{j_1} > \left(\frac{m}{M}\right)^{1/2}. \tag{5.14}$$

In the general case,

$$\frac{j_+}{j_1} \geqslant \left(\frac{m}{M}\right)^{1/2}. \tag{5.15}$$

To be more precise,

$$\frac{j_+}{j_1} \geqslant \gamma\left(\frac{m}{M}\right)^{1/2} \tag{5.16}$$

where the coefficient γ is from 3/2 to 3, depending on the state of surface of the cathode [5.14].

Note that the current densities j_1 and j_+, determined by equations (5.11) and (5.12), are approximately twice as large as those in the vacuum diode, owing to the presence of oppositely charged particles and some neutralization of the space charge by these particles. According to Bohm' theorem [5.13], inequality (5.15) is the necessary condition of stationarity of the layer at the cathode: if it is not satisfied, there is no stationary solution to Poisson's equation. In fact, the moment of ignition of the discharge is defined as the moment of formation of the cathode layer: the current of primary electrons then increases to the value given by (5.11), in accordance with the decrease in d (of course, it is true for the case when the temperature of the cathode is sufficiently high for providing the required emission current). To ignite the discharge, the plasma must send to the boundary of the layer an ion flux given by (5.12),

$$j_+ = en_+v_+ \geqslant j_1\left(\frac{m}{M}\right)^{1/2} \tag{5.17}$$

where, as we could see in section 5.2, $n_+ = n_e e^{-1/2}$ and $v_+ = (T_e/M)^{1/2}$. In fact, the physics of the phenomena is generally the same as when ions are collected by a negatively charged probe: indeed, by virtue of Bohm's theorem, the entire probe current is also determined by the electron temperature, namely [5.13],

$$j_+ = en_e e^{-1/2} \left(\frac{T_e}{M}\right)^{1/2} \tag{5.18}$$

or, which is an equivalent expression,

$$j_+ \simeq 0.4en_e \left(\frac{2T_e}{M}\right)^{1/2}. \tag{5.19}$$

Relations (5.15) and (5.17) give

$$j_1 \lesssim en_e v_e/4 \equiv j_e \tag{5.20}$$

or

$$n_1 u \lesssim n_e v_e/4. \tag{5.21}$$

The last formula is quite lucid: the current density of primary beam electrons in a stable discharge with a hot cathode (uniform along its length) cannot exceed the density of the random current of plasma electrons†.

Condition (5.15) of discharge ignition requires that the density n_0 of the neutral gas in the discharge chamber exceed a certain threshold. Indeed, the rate of ionization in a plasma column is always proportional to n_0; for example, the rate of ionization by primary electrons per unit time and unit area of column cross section is

$$eN_+ = j_1 n_0 \sigma_i L \tag{5.22}$$

where L is the gas column length and σ_i is the ionization cross section. If this mechanism of ion formation is predominant, and the strong magnetic field 'allows' ions in the beam to move only along the beam, then (5.15) and (5.22) imply that

$$j_+ \simeq \frac{eN_+}{2} = \frac{1}{2} j_1 n_0 \sigma_i L > \left(\frac{m}{M}\right)^{1/2} j_1$$

that is,

$$n_0 > \frac{2}{\sigma_i L} \left(\frac{m}{M}\right)^{1/2}. \tag{5.23}$$

Hence, the ignition of the discharge is a threshold effect: as the gas pressure in the discharge chamber is gradually increased, a moment comes when the

---

† It is important to note, in view of topics that follow, that the value of j_e is determined by the electron density n_e in the neighbourhood of the cathode layer.

cathode layer is formed and the beam current and plasma density increase jumpwise. (The reader shall recall that the plasma–beam discharge in the version that was meant in section 5.2 ignites also at a constant beam current, by a beam generated using an autonomous electron gun whose operation is independent of the charge density in the newly generated plasma.)

So far I have discussed only the left-hand (cathode) region of the curve of potential distribution along the plasma column. Now we can turn to the right-hand (anode) region (here I refer to the beam collector as the anode). If the discharge is maintained in a longitudinal magnetic field, then, as was already clear in the derivation of relation (5.7), a negative anodic potential drop φ sets in at the anode: the plasma potential here is higher than that of the anode. This excess is caused by the requirement of quasineutrality of the plasma ($n_e = n_+$), since the electron and ion fluxes from the plasma in a stationary state must be equal:

$$2n_+ \exp(-1/2)\left(\frac{T_e}{M}\right)^{1/2} = \frac{n_e v_e}{4} \exp\left(-\frac{e\varphi}{T_e}\right). \qquad (5.24)$$

The coefficient 2 in (5.24) appears since ions escape both to the cathode and to the anode. Equation (5.24) implies that

$$\varphi \simeq \frac{T_e}{e} \ln\left(\frac{M}{9m}\right)^{1/2}. \qquad (5.25)$$

If the plasma beam length greatly exceeds the length of the discharge chamber (into which the gas is fed) and ions are generated almost exclusively in the discharge chamber (see below), then a longitudinal plasma density gradient is formed in the beam: the plasma density in the main part of the plasma beam is considerably lower than in the discharge chamber close to the cathode layer. Near the discharge ignition threshold, therefore, the ratio of the current density of the primary electrons to the density of the random current of plasma electrons in the entire plasma beam, except in the region in the discharge chamber where the beam is generated, is

$$j_e \equiv \frac{n_e v_e}{4} < j_1 \equiv n_1 u. \qquad (5.26)$$

The difference between (5.26) and (5.20) need not seem surprising: it is due to plasma inhomogeneity along the discharge column.

The choice of equipment used in experiments with plasma beams is dictated by the choice of discharge regime. The plasma source [5.15] shown in figure 5.2 proves to be quite suitable for stationary discharge. To generate high-density plasma (10^{13} cm^{-3}), a pulsed high-current-density discharge is used, for which the discharge chamber has a large reserve volume shown in figure 5.3. The need to have this volume is caused by the fact

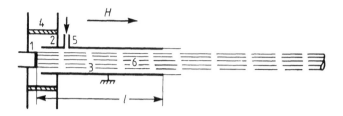

Figure 5.2 Schematic arrangement of a plasma source for continuous discharge: 1, tungsten cathode (heated by electron bombardment); 2, diaphragm; 3, discharge chamber; 4, insulator; 5, gas inlet; 6, plasma beam; $l = 15$ cm.

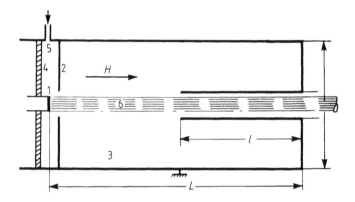

Figure 5.3 Schematic arrangement of a plasma source for pulsed discharge; $l = 25$ cm, $L = 35$ cm. Notation is the same as in figure 5.2.

that the real gas pressure in the discharge chamber is determined by the balance of the inflow of the gas and the 'pumping' of particles out of the chamber. When the discharge is ignited, this pumping is enhanced since ions possess greater velocities than neutral molecules. If the discharge is sufficiently intense and the volume of the discharge chamber is insufficient, the pressure in the discharge chamber falls below the discharge ignition threshold, and the discharge stops. After a certain time, the gas pressure grows again, reaches the ignition threshold, the discharge starts up again, and so forth. As a result, the discharge regime is relaxational. The reserve volume shown in figure 5.3 serves precisely the purpose of increasing the discharge pulse duration. This design of the plasma source makes it possible to sustain a discharge of current density of, for example, $50 \, \text{A/cm}^2$ with pulse length of 50 ms, in contrast to the design of figure 5.3, in which the discharge regime becomes relaxational already at discharge current densities of several amperes per $1 \, \text{cm}^2$ (with pulse length of $200 \, \mu\text{s}$ and pauses on the order of 1 ms).

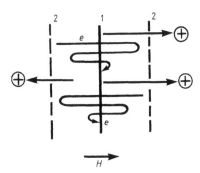

Figure 5.4 Schematic diagram of a high-current pulsed ion diode (triode): 1, cathodes; 2, anode. Arrows indicate the directions of motion of electrons and ions.

I will remark in conclusion that a discharge with a hot cathode is a simple and convenient method of generating a high-density plasma beam with readily controlled charged particle density up to 10^{12}–$10^{13}\,\mathrm{cm^{-3}}$. The discharge with a hot cathode can be called a variation of the plasma–beam discharge, whose properties are treated additionally in section 6.11.

5.5   High-Power Pulsed Ion Beams

Let us look at the so-called reflex discharge, or the Penning discharge, which is maintained along a strong external magnetic field in a three-electrode system of two cathodes and one anode (the anode is transparent for particles) (figure 5.3). In this design, the primary electrons emitted from one of the cathodes and accelerated towards the anode do not reach the opposite cathode, owing to an unavoidable loss of some energy during one time of flight; hence, they perform a large number of flight oscillations until they are deposited at the anode. As a result, the amount of primary electrons crossing the plane of the anode (in both directions) is much greater than the number of these particles that leave the cathodes during the same time. The space charge of electrons in the neighbourhood of the anode is correspondingly considerably higher than in the case of the direct (oscillation-free) discharge between the cathodes and the anode (for the same flux of electrons leaving the cathode). An additional important cause of increase in the space charge of electrons at the anode (as compared with the direct discharge) is the quite significant slowing down of electrons before they are deposited on the anode. For these reasons, the ion current to the cathodes is much greater than the quantity given by expression (5.13). If the cathodes (or at least one of them) are made sufficiently transparent for ions, a highly efficient ion source is obtained: its efficiency (the ratio of the powers of the ion fluxes to the power of the electron beams) should be much greater than $(m/M)^{1/2}$ (which is the case for the direct discharge)†. Modern high-voltage pulsed ion diodes operate on this principle; using them, it was possible to generate ion beams with unexpectedly impressive parameters: current up to hundreds of kiloamperes, energy of ions up to hundreds of keV, efficiency of 40% or higher, pulse lengths of the order of 50–100 ns [3.28, 5.18–5.21]. The recent dramatic progress of ion diodes should be noted and one can be sure that their parameters will greatly surpass these values in the not too distant future.

5.6   Measurement of Space Potentials, Electric Fields and Particle Energies and Concentration in Beam Plasma

5.6.1   Differential Anode
A familiar method of determining the plasma density from the measurement of ion current to a probe whose potential relative to the plasma is sufficiently negative is the Langmuir probe method [5.13]. The presence of an intense beam and the ensuing possibility of destruction of the probe create considerable obstacles for the application of this technique. In this situation, a very efficient method is the method of the so-called differential anode [5.22–5.24] (figure 5.5). A small hole is made in the beam receptacle (the anode), with a sufficiently massive collector mounted on insulators behind the hole. The current–voltage characteristic of the collector (whose potential is measured relative to the anode) is used as the actual probe characteristic from which the information on plasma density is extracted: according to Langmuir's formula (5.18) (see [5.14]) elaborated by Bohm [5.13], the density of the probe saturation ion current is [5.19]:

$$j_+ = 0.4en_+\left(\frac{2T_e}{M}\right)^{1/2}.$$

Figure 5.6 shows a real probe characteristic of a differential anode (reproduced from [5.25]).

5.6.2   Method of Retarding Field
The differential anode (see section 5.6.1) can be used to analyse the energy distribution of primary beam electrons by the retarding field method [5.26, 5.27]. However, this means that effects connected with the secondary emission of electrons knocked out by the primary beam from the collector and the anode must be eliminated. If the density of plasma electrons in the beam is low, the secondary emission is efficiently suppressed by introducing

---

† In real situations, one of the cathodes may be the virtual cathode formed when the primary electron beam is shut off by its own space charge.

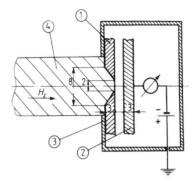

Figure 5.5 Schematic of a differential anode: 1, anode (tantalum); 2, collector (tantalum); 3, screen; 4, beam. Arrows indicate the size in mm. The spacing between the electrodes 1 and 2 is 1.5 mm.

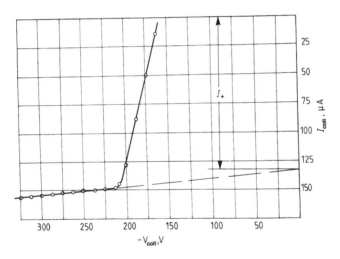

Figure 5.6 Ion branch of the current–voltage characteristic of the differential anode placed at the centre of the plasma beam: the collector ion current as a function of the (negative) collector potential.

a sufficiently transparent grid of thin wire into the space between the anode and the collector and by connecting it to a negative potential (relative to the collector) of 25–50 V. If, however, we deal with a plasma beam in which the density of plasma electrons is (considerably) greater than that of the primary electrons, then the 'grid' must perform another function: to destroy the plasma which 'flows' into the hole in the differential anode. To achieve this, it is sufficient to shape the grid as a relatively thick plate with a hole whose diameter must be greater than the diameter of the inlet hole in the

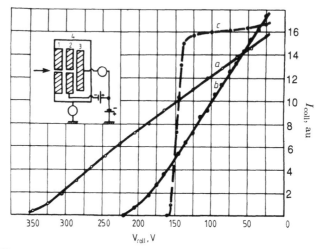

Figure 5.7 Energy spectrum analyser for fast electrons of the plasma beam [5.28]: 1, discharge anode; 2, 'grid'; 3, collector; 4, screen. The arrow shows the direction of beam propagation; (*a*), (*b*) and (*c*), examples of collector current versus voltage characteristics: (*a*) and (*b*), in the plasma–beam discharge with the beam electron energy of $W_1 = 300\,\text{eV}$ and $W_1 = 160\,\text{eV}$, respectively, and (*c*), in a vacuum, $W_1 = 160\,\text{eV}$.

differential anode. This arrangement was used in the experiments of [5.28] on the analysis of stability of a high-density plasma beam (figure 5.7). The energy distribution of the particles to be analysed is determined from the retarding current–voltage characteristic.

In experiments on accelerating and heating the ions in plasma beams (see chapter 8), two simple analyser probes were used to measure, by the retarding field method, the energy of ions moving both at right angles to the external magnetic field and along this field. Their design is shown schematically in figures 5.8 and 5.9. Protons whose Larmor rotation energy was above several tens of eV reached the probe collector (figure 5.8). The presence of a (transverse) magnetic field of several Oe eliminated the need to suppress the secondary electron emission. The ion energies were found from the current–voltage characteristic of the collector. A similar but smaller-size probe was used to evaluate the density of fast ions. The probe for measuring the energies of motion of ions along the external magnetic field (figure 5.9) consisted of five electrodes: four molybdenum grids and a collector. The energies of ions were found from the dependence of the ion current to the collector on the potential V_3 of the third (control) grid. Using these probes, energy spectra of protons in the keV energy range were measured (see sections 6.4 and 7.12).

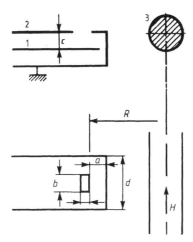

Figure 5.8 Analyser probe for ion rotation energy in magnetic field (in two projections) [5.34]: 1, collector; 2, screen; 3, plasma beam.

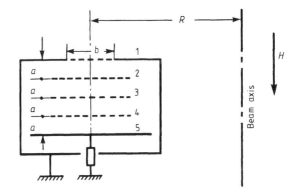

Figure 5.9 Analyser probe for the energy of ions moving along the magnetic field [5.15]: 1, screen and grid, 2–4, grids; 5, collector. Grid 3 is the control grid; $a = 5$ mm, $b = 10$ mm.

5.6.3   Reverse Current Probe: an Indicator of a Virtual Cathode in Electron Beams and a Virtual Anode in Ion Beams

An ingenious version of the retarding field technique is the method of 'reverse current' analysis, suggested in [5.28] and applied to study the phenomenon of shutting off the originally neutralized electron beam by its own space charge, when the beam current exceeds a certain limiting value. In this method, one measures the energies of electrons moving against the beam direction: when the beam shuts off, the energy of the 'counter-streaming' electrons are obviously the same as the initial electron energies in the non-perturbed beam. The energies of oppositely moving

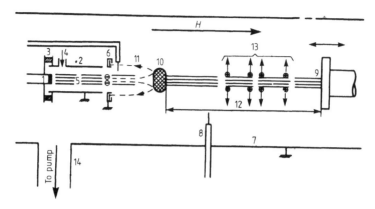

Figure 5.10   Reverse current probe 6 for measuring the energy spectrum of electrons moving against the direction of propagation. of the plasma beam [5.28, 5.29]. The gap between the beam and the probe is not greater than 1–1.5 mm: 1, cathode, 1 cm in diameter, at a negative potential V_d; 2, discharge chamber of 2 cm in diameter, 15 cm long, at zero potential; 3, insulator; 4, gas inlet; 5, beam; 7, vacuum chamber; 8, probes; 9, anode; 10, virtual cathode; 11, trajectories of beam electrons at the moment the virtual cathode is formed; 12, region of positive potential; 13, accelerated ions; 14, pumping.

electrons are measured by the retarding field method in its most miniature implementation. Namely, a two-electrode analyser is used; its working electrode is a ring surrounding the beam; its plane is perpendicular to the velocity of beam electrons. The ring is screened by a metal envelope (the second electrode of the analyser) on all sides, excluding a narrow slit into which only oppositely directed electrons can move (figure 5.10). This device is an unambiguous indicator of the formation of a virtual cathode in the electron beam. The following clarification is needed here, however. The beam electrons reflected from the virtual cathode can penetrate the working slit of the analyser in question only if they manage to move radially (transversely to the magnetic field H) by a distance $\Delta r \gtrsim 2$ mm. This will not happen if the electric field E of the space charge of the beam possesses axial symmetry. Indeed, the maximum radial displacement of a beam electron along the field E, transversely to the magnetic field H, is then equal to the length of the cycloid arc

$$a = 2mc^2 E/eH^2$$

and will be less than 0.1 mm at $E \simeq (2-3)$ CGSE and $H \gtrsim 10^3$ Oe. Hence, the device we discuss could work only if the azimuthal component E_φ of electric field was sufficiently strong. The radial displacement of an electron in crossed fields E_φ and H_z is

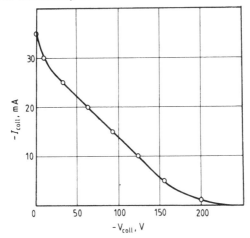

Figure 5.11 Current–voltage characteristic of a 'reverse current probe' in a specific regime of beam–plasma discharge (at a sufficiently low gas pressure in the discharge chamber of a plasma source and, consequently, low plasma density). The beam electron energy at the input to the system was 200 eV. Beam electrons are seen to be reflected from the virtual cathode [5.29].

$$\Delta r = c \, \frac{E_\varphi}{H_z} \frac{L}{u}$$

where L is the distance between the virtual cathode and the analyser; if $E_\varphi \simeq (2-3)\,\mathrm{CGSE}$, $L \simeq 10\,\mathrm{cm}$, $u = 10^9\,\mathrm{cm/s}$ and $H_z = 10^3\,\mathrm{Oe}$, this relation gives $\Delta r \simeq 1\,\mathrm{cm}$, which is much greater than the indicated sensitivity threshold of the method. I shall point out, even at the cost of running ahead of the presentation a little, that electric fields with the component E_φ of this and higher value are indeed observed in self-shutting neutralized beams (see chapter 6). The device described above is often referred to for the sake of brevity as a *reverse current probe*. The results of applying a reverse current probe to studying the current shut-off in electron and plasma beams are described in chapter 6. In the simplest version of this method, the retarding current–voltage characteristic of the type shown in figure 5.11 is not recorded; one simply measures the 'floating' potential to which is charged a well-insulated reverse-current probe collector (see [5.29]). This potential, just as the maximal electron energy in figure 5.11, characterizes unambiguously the energy of electrons moving against the beam.

5.6.4 Measurement of Spatial Potential and Electric Fields in Ion Beams by Measuring the Energy of Secondary Ions

Let us now look at another version of applying the retarding field method, which is very suitable (in a number of experimental situations) not only for

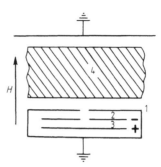

Figure 5.12   Three-electrode energy analyser for secondary
ions leaking from a quasineutral ion beam [5.30, 31]. 1, entrance
electrode; 2, intermediate electrode ('grid'); 3, ion collector; 4,
beam.

an energy analysis of particles but also for measuring the spatial potential
and electric fields in beams. The method was suggested and then elaborated
in [5.30–5.32].

Imagine an ion beam propagating along the axis x (perpendicularly to
the plane in figure 5.12) and having a sufficiently large size along the axis
y. The beam passes through the plasma containing secondary (slow) ions.
If the beam carries an electric field with a component $E_z \neq 0$ (the z axis
is directed along the magnetic field), the secondary ions that leak from the
beam along the direction of z arrive at the walls of the set-up with energies
that are practically equal to the potential difference that they traverse in
the field E_z. By measuring the ion energies, we obtain the potential drop
in the ion beam along z and, knowing the beam size along z, we find the
field strength E_z. Having carried out such measurements at points with
different coordinates y and x, we obtain a complete picture of the potential
distribution along the axes y, x, that is, the distribution of electric fields
in the beam.

The method of retarding field, in the version [5.30, 5.31], which has
now become widespread [5.27], has been used in the work described above.
The experimental set-up [5.30, 5.31] was placed horizontally under the
ion beam at the midpoint of its trajectory (where the beam width, i.e.,
its size in the y direction, was several tens of centimetres) and included
19 measuring cells at different coordinates y (see figure 5.12). Each cell
consisted of a collector and two diaphragms with slits 1 mm wide (along
the y axis) and 5 cm long (along the x axis, i.e., along the beam); the
collectors were 10 cm long and 2 cm wide; the cells were spaced by 3 cm. To
single out the ion current, a negative potential of about 75–100 V (relative
to the grounded input diaphragm) was applied to the second (intermediate)
analyser diaphragm. After this, the retarding characteristic was recorded:

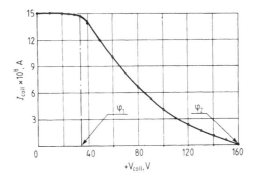

Figure 5.13 Typical example of integral energy spectrum of secondary ions (N_2^+) leaking from a quasineutral beam of Zn^+ ions; beam ion energy 30 keV and beam current 100 mA [5.30, 31].

the collector ion current as a function of collector potential (figure 5.13). The minimal ion energy in this case is about 40 eV. This value characterizes the minimal beam potential at the boundary with the secondary plasma surrounding it.

The maximal ion energy in figure 5.13 is 160 eV; it characterizes the maximum potential in the beam for given x and y. With the known distribution of the quantities φ_1 and φ_2 across the beam (i.e., along y), it is possible to find the vertical (along z) and horizontal (along y) components of the electric field in the beam,

$$E_z \simeq \frac{\varphi_2 - \varphi_1}{h/2} \qquad E_y \simeq \frac{\Delta\varphi_2}{\Delta y}$$

where h is the beam height. For instance, if $\varphi_2 - \varphi_1 = 120\,V$ (see figure 5.13) and $h = 5\,cm$, we find $E_z \simeq 50\,V/cm$.

Obviously, an indicator of the beam potential and of electric fields in the beam can be—and often is—the 'floating' potential of a (carefully insulated) analyser collector. Figure 5.14 gives specific examples of potential distribution in a 40 keV Cd^+ ion beam with a current of hundreds of milliamperes, which propagates through the vacuum and carries strong (a) and weak (b) electric fields. The application of the retarding field method to measure ion energies in quasineutral electron beams, plasma beams and magnetic traps is described in sections 6.4, 7.1 and 7.2.

Other types of analysers are also applied to measure ion energy distributions [5.33].

5.6.5   Analysis of Fast Neutral Charge-Exchange Atoms

The analysis of fast neutral atoms ejected from the plasma is an efficient method of measuring the energy distribution of ions. Fast neutral atoms

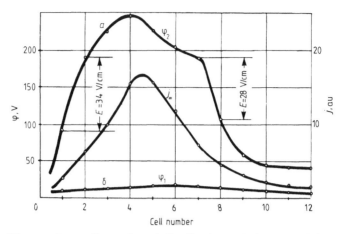

Figure 5.14 Examples of distributions of the potential φ and the DC component of the current density $j_=$ (transversely to the beam of Cd$^+$ ions), for various degrees of neutralization of its space charge [5.30, 5.31].

are formed as a result of charge exchange of accelerated ions in the residual gas of the plasma. In the analyser, they are again converted back into ions (while passing through a thin foil or a special charge exchange chamber), after which the energy spectrum of ions is obtained by an electrostatic technique (e.g., by the Hughes–Rojanski method [5.33]).

The measurement of charge-exchange neutral atom flux is also used as a method of determining the density of fast ions in the plasma [5.15, 5.34, 5.35]. A device realizing this is a sensor consisting of two insulated plates (of stainless steel), slightly tilted towards the central plain of the set-up (this plane is perpendicular to the beam direction). Secondary emission electrons knocked out by fast neutral atoms from one plate are collected at the other; their current density j_n is a measure of the density of fast ions in the plasma. The coefficient of the ion–electron secondary emission is usually assumed to equal unity (at ion energies $\sim 1\,\text{keV}$) [5.35]. The fast ion density n_+ is given by the expression

$$j_n = n_0 n_+ \langle \sigma v \rangle_n \frac{\pi R_{\text{av}}^2}{2\pi R_0}$$

where n_0 is the density of molecules of the neutral gas, σ_n is the effective cross section of charge exchange of ions moving in a neutral gas at a velocity v, R_0 is the radius at which the sensor is located, R_{av} is the average radius of the plasma containing fast ions. The quantity $\langle \sigma v_n \rangle$ is evaluated from the available experimental data (e.g, see [5.36]) for ion energies measured by the analyser (e.g., of the type shown in figure 5.8). Thus if the hydrogen ion energy is $1\,\text{keV}$, we have $\langle \sigma v \rangle_n \simeq 2.4 \times 10^{-8}\,\text{cm}^3\,\text{s}^{-1}$.

5.6.6   Probing of Electric Fields by Electron and Ion Beams

So far, we have discussed passive methods of particle diagnostics. The present section is devoted to active techniques. One of the modern methods of measuring electric fields in the plasma and in the beam is to measure the deviation displayed by a special 'probing' charged particle beam. This may be an electron or an ion beam. A probing beam is quite convenient if field variations are relatively slow. In these conditions, an ion beam is an even more universal probe than an electron beam since it can be used to shoot through the plasma in a magnetic field, sending the beam not only along but also at right angles to this field. An electron beam is suitable for transverse probing of a plasma only if the magnetic field is very weak or absent altogether.

In [5.37], an electron probe beam was sent along the axis of a hollow cylindrical electron beam in order to determine the non-neutralized space charge of this beam. The space charge of the beam to be analysed produced, under certain conditions, such a considerable depression of potential along the axis of the system that the probing beam was reflected from the electrostatic 'mirror' thus formed and could not reach the collector. The threshold after which the probing beam gets to the collector can be measured by increasing the energy of this beam; this energy is equal to the depression of the potential, caused by the space charge of the main beam.

The method applied in [5.38] was also the transverse probing of the plasma by an electron beam; an improvement was introduced, which made it possible to measure not only the field (by recording the deflection of the probing beam) but also its oscillation frequency. To carry out such measurements, the probing beam, transmitted through the plasma volume to be studied, was sent through a capacitor whose plates were connected to a special deflection voltage generator of variable frequency and amplitude; the electric field in this capacitor is perpendicular to the field under consideration. Having passed through the capacitor, the probing beam hits a luminescent screen on which it traces generally quite complicated trajectories. If the oscillator frequency is properly selected, Lissajous patterns are obtained on the screen, giving the frequency at which the field oscillates.

The same method was applied successfully to identify Langmuir oscillations in experiments with the so-called *cavitons* [5.39], that is, wave bunches localized within a short segment of the plasma column, and to measure the magnitude of the electric field of these oscillations. It was possible to apply the method of transverse electron beam probing because no longitudinal magnetic field was present in these experiments.

5.6.7   The Capacitive Probe Method

The capacitive method used in [5.40] for an experimental study of the

dynamics of current shut-off in a non-neutralized electron beam also belongs
to the group of methods of passive diagnostics. It is based on the fact that
the electric field of a non-neutralized beam (the radial field in the case of
the geometry like that in figure 5.2) induces electric charges on the plates
of a special diagnostic capacitor. The receiving electrode of this capacitor is
a ring surrounding the beam and placed close to the wall of the cylinder D
surrounding the beam and acting as the second electrode of the capacitor.
The electric field E created by the beam close to the surface D is related
to the charge q per unit length of the beam:

$$E = 2q/R.$$

If a beam of current I, propagating at a velocity u, is not neutralized by ions
at all, the voltage difference across the plates of the diagnostic capacitor is

$$\Delta\varphi = Ed = 2Id/uR$$

where d is the gap between the plates. This capacitive method is convenient
for analysing the electric field of a 'bare' beam. If, however, the beam
propagates through a plasma, it is necessary to protect the probe from
plasma particles, namely, to make a thin-wire receiving electrode. The
probe can thus be made 'transparent' for particles but 'opaque' for the
electric field.

5.6.8   Measurement of Electric Fields of Waves in Beam Plasma by Spectroscopic Methods

Spectroscopic methods of measuring electric fields in the beam plasma
involve, first of all, measuring the Stark broadening of the spectral lines
of atoms (or ions) [5.41] and, second, measuring the intensity of forbidden
satellites of spectral lines [5.41–5.47]. Thus electric fields up to 1.5 MeV/cm
were measured in intense ion beams in the experiments [5.41]. The second
of these methods is based on the fact that if the frequency of oscillations in
the field E to be studied is close to the difference between the frequencies
of the allowed and the (neighbouring) forbidden lines, the forbidden line
(the satellite) should be observed in the field E and its intensity should
grow steeply with the field strength E. In [5.46], both these methods were
used to study important characteristics of of the interaction between the
ion flow and the plasma, and in [5.44] the basic data on the magnitude
and the spatial structure of the electric fields of plasma waves in the
regime of strong Langmuir turbulence were obtained by measuring the
intensity of the forbidden satellites. In a similar manner, electric fields of
Langmuir waves ($E \simeq 28\,\mathrm{kV/cm}$) were measured in high-density plasma
($n_e \simeq 5 \times 10^{13}\,\mathrm{cm}^{-3}$) produced in the gas by a high-intensity relativistic
electron beam [5.45].

5.6.9   Measurement of Electric Fields in Beam Plasma by the Method of Collective Scattering of Laser and Microwave Radiation

The methods indicated in the title of this section are based on using Thomson scattering of light by electrons [5.48, 5.49] for studying the collective (wave) motions of plasma particles. If the wavelength of the plasma oscillations is considerably greater than the Debye radius, the scattering of the radiation by the electrons participating in the oscillations is found to be collective. The intensity of the scattered radiation therefore grows by many orders of magnitude, in proportion to the squared amplitude of oscillations of plasma density. If scattering is collective, then the conservation of energy and momentum of quanta imply the following relations between the frequencies and wave vectors of the incident wave (ω_i, k_i), of the scattered wave (ω_s, k_s) and of the plasma wave (ω, k):

$$\omega_s = \omega_i \pm \omega \quad k_s = k_i \pm k.$$

In other words, the scattering changes the radiation frequency by

$$|\Delta\omega| = \omega$$

and the wave vector changes by $\Delta k = k_s - k_i$. The absolute value of the change in the wave vector at small scattering angles is

$$|\Delta k| = 2k_i \sin(\Theta/2)$$

where Θ is the angle by which the incident wave is scattered. The collective scattering is thus 'resonant'-amplified at those angles Θ at which $\Delta k \simeq k$, that is,

$$k = 2k_i \sin(\Theta/2)$$

or

$$\lambda_i = 2\lambda \sin(\Theta/2)$$

where $\lambda_i = 2\pi/k_i$, $\lambda = 2\pi/k$.

This last relation is the familiar Bragg–Wulf formula for the scattering of radiation by an ordered structure. In this particular case, the structure is the Fourier harmonic of the spectrum of the oscillations under considerations with a specific wave number k and wavelength $\lambda = 2\pi/k$. If we measure the power distribution of the scattered radiation over angle Θ, then it is easy to determine from the above relations the wavelength and frequency spectra of plasma oscillations. Such measurements immediately give the dispersion relation, $\omega(k)$, for the waves. The main characteristics of electron and ion waves in plasma beams were measured using this method [5.50–5.57].

Even though this type of scattering is caused by coherent contributions of oscillating plasma particles (electrons), it is known in a large part of the literature as non-coherent (although collective!) scattering, since it is accompanied by a change in frequency. Another equivalent name for this phenomenon is Raman scattering†.

One of the more recent efforts to study the spectrum of drift waves in a large Tokamak by the method of collective scattering of infrared laser radiation ($\lambda = 337\,\mu$m) was described in [5.58].

5.6.10   Plasmascope Investigation of Transverse Structure of (Beam) Plasmas

A very original and efficient plasmascope diagnostic was developed in [5.59–5.61], enabling one to photograph the transverse structure of the plasma on a fluorescent screen whose light emission is caused by plasma particles leaking to one of the end faces of the discharge column‡.

The plasmascope method gave extremely clear photographs of plasma 'protuberances' that emerge from the plasma column as a consequence of complex processes which are sometimes referred to (very imprecisely) as 'plasma diffusion in magnetic field'.

References

[5.1]   Getty W D and Smullin L D 1963 *J. Appl. Phys.* **34** 3421

[5.2]   Kornilov E A, Kovpik O F, Fainberg Ya B and Kharchenko I F 1965 *Zh. Tekhn. Fiz.* **35** 1372, 1378

[5.3]   Lebedev P M, Onishchenko I N, Tkach Yu V, Fainberg Ya B and Shevchenko V I 1976 *Fizika Plazmy* **2** 407 (Engl. Transl. 1976 *Sov. J. Plasma Phys.* **2** 222)

[5.4]   Antipov S V, Nezlin M V, Snezhkin E N and Trubnikov A S 1973 *Zh. Eksp. Theor. Fiz.* **65** 1866 (Engl. Transl. 1974 *Sov. Phys.-JETP* **38** 931)

[5.5]   Snezhkin E N and Nezlin M V 1977 *Zh. Eksp. Theor. Fiz.* **73** 913 (Engl. Transl. 1977 *Sov. Phys.-JETP* **46** 481)

[5.6]   Fainberg Ya B, Shapiro V D and Shevchenko V I 1969 *Zh. Eksp. Theor. Fiz.* **57** 966 (Engl. Transl. 1970 *Sov. Phys.-JETP* **3** 528)

[5.7]   Gorbatenko M F and Shapiro V D 1965 In *Interaction of Charged Particle Beams with Plasma* (Kiev: Naukova Dumka) p 103 (in Russian)

[5.8]   Kellog P J and Boswell R W 1986 *Phys. Fluids* **29** 1669

† See chapter 11.

‡ The plasma was produced in the experiments [5.59–5.61] by an electron beam propagating through the gas along the magnetic field.

[5.9]   Ivanov A A and Soboleva T K 1978 *Non-Equilibrium Plasmachem-istry* (Moscow: Atomizdat) (in Russian)

[5.10] Kovalenko V P 1983 *Uspekhi Fiz. Nauk* **139** 224 (Engl. Transl.1983 *Sov. Phys.-USPEKHI* **26** 116)

[5.11] Lavrovsky V A, Kharchenko I F and Shustin E G 1973 *Zh. Eksp. Theor. Fiz.* **65** 2236

[5.12] Gadeev K K, Erastov E M, Ivanov A A, Muksunov A M, Nikiforov V A, Severnyi V V, Khripunov B I and Shapkin V V 1981 *Dokl. AN SSSR* **256** 834

[]   Popkov N G 1990 *Physika Plasmy* **16** 1375

[5.13] Bohm D 1949 *The Characteristics of Electrical Discharges in Magnetic Fields* ed A Guthrie and R K Wakerling (New York: McGraw Hill)

[5.14] Langmuir I 1929 *Phys. Rev.* **33** 954

[5.15] Nezlin M V and Solntsev A M 1965 *Zh. Eksp. Theor. Fiz.* **48** 1237 (Engl. Transl. 1965 *Sov. Phys.-JETP* **21** 826)

[5.16] Gabovich M D 1972 *Physics and Technology of Ion Plasma Sources* (Moscow: Atomizdat) (in Russian)

[5.17] Gabovich M D, Pleshivtsev N V and Semashko N N 1986 *Ion and Atom Beams for Controlled Thermonuclear Fusion and Technological Applications* (Moscow: Energoatomizdat) (in Russian)

[5.18] Olson C L 1977 *Fizika Plasmy* **3** 465 (Engl. Transl. 1977 *Sov. J. Plasma Phys.* **3** 259)

[5.19] Bystritsky V M and Didenko A N 1984 *High-Power Ion Beams* Moscow: Energoatomizdat) (in Russian)

[5.20] Humphries S, Jr 1980 *Nucl. Fusion* **20** 1549

[5.21] Olson C L 1982 *J. Fusion Energy* **1** 309

[5.22] Tonks L, Mott-Smith H M and Langmuir I 1926 *Phys. Rev.* **28** 104

[5.23] Kovalenko V F, Rozhansky D A and Sena L A 1934 *Zh. Tekhn. Fiz.* **4** 1688

[5.24] Zharinov A V 1959 *Atomnaya Energiya* **7** 215 (Engl. Transl. 1961 *Sov. J. At. Energy* **7** 710)

[5.25] Nezlin M V 1961 *Zh. Eksp. Theor. Fiz.* **41** 1015 (Engl. Transl. 1962 *Sov. Phys.-JETP* **14** 723)

[5.26] Ionov N I 1952 *DAN SSSR* **85** 753

[5.27] Ionov N I 1965 *Zh. Tekhn. Fiz.* **34** 769

[5.28] Nezlin M V 1964 *Zh. Eksp. Theor. Fiz.* **46** 36 (Engl. Transl. 1964 *Sov. Phys.-JETP* **19** 26)

[5.29] Nezlin M V and Solntsev A M 1965 *Zh. Eksp. Theor. Fiz.* **49** 1377 (Engl. Transl. 1966 *Sov. Phys.-JETP* **22** 949)

[5.30] Nezlin M V 1960 *Zh. Tekhn. Fiz.* **30** 168 (Engl. Transl. 1960 *Sov. Phys. Tech. Phys.* **5** 154)

[5.31] Nezlin M V 1967 *Plasma Phys.* **10** 387

[5.32]   Nezlin M V and Morozov P M 1959 *Proceedings of the 3rd Intern.*
         Conf. on Peaceful Uses of Atomic Energy Geneva-58 **6** (Moscow:
         Atomizdat) p 117

[5.33]   Afanasiev V P and Yavor S Ya 1978 *Electrostatic Analyzers for*
         Charged Particle Beams (Moscow: Nauka) (in Russian)

[5.34]   Nezlin M V and Solntsev A M 1963 *Zh. Eksp. Theor. Fiz.* **45** 840
         (Engl. Transl. 1964 *Sov. Phys.-JETP* **18** 576)

[5.35]   Ioffe M S, Sobolev R I, Telkovsky V G and Yushmanov E E 1960 *Zh.*
         Eksp. Theor. Fiz. **39** 1602 (Engl. Transl. 1961 *Sov. Phys.-JETP*
         12 1117)

[5.36]   Fite W L, Brackman R T and Snow W R 1958 *Phys. Rev.* **112**
         1161

[5.37]   Fedorchenko V D, Muratov V I and Rutkevich B M 1964 *Zh. Tekhn.*
         Fiz. **34** 463

[5.38]   Gabovich M D, Gladkii A M, Kovalenko V P *et al* 1973 *Pis'ma v*
         Zh. Eksp. Teor. Fiz. **18** 343 (Engl. Transl. 1973 *Sov. Phys.-JETP*
         18 202)

[5.39]   Kim H C, Stenzel R and Wong A Y 1974 *Phys. Rev. Lett.* **33** 886

[5.40]   Volosok V I and Chirikov B V 1957 *Zh. Tekhn. Fiz.* **27** 2624 (Engl.
         Transl. 1957 *Sov. Phys. Tech. Phys.* **2** 2437)

[5.41]   Maron Y, Coleman M D, Hammer D A and Peng H-S 1986 *Phys.*
         Rev. Lett. **57** 699

[5.42]   Amini B 1986 *Phys. Fluids* **29** 3775

[5.43]   Kuznetsov E I and Shcheglov D A 1980 *Methods of Diagnostics of*
         High-Temperature Plasma 2nd edition (Moscow: Atomizdat)

[5.44]   Eremin B G, Kostrov A V, Lunin N V and Stepanushkin A D
         1977 *Izv. VUZov, Ser Radiofizika* **20** 1489 (Engl. Transl. 1978
         Radiophys. Quantum Electron. **20** 1025)

[5.45]   Dovrat A and Benford G 1989 *Phys. Fluids* **1B** 2488

[5.46]   Pistunovich V I, Platonov V V, Ryutov V D and Filimonova E A
         1976 *Pis'ma v Zh. Eksp. Teor. Fiz.* **23** 30 (Engl. Transl. 1976
         JETP-Lett. **23** 26)

[5.47]   Levron D, Benford G, Baranga A and Means S 1988 *Phys. Fluids*
         31 2026

[5.48]   Pyatnitsky L N 1976 *Laser Diagnostics of Plasmas* (Moscow:
         Atomizdat) (in Russian)

[5.49]   Lukyanov S Yu 1975 *Hot Plasmas and Controlled Fusion* (Moscow:
         Nauka)

[5.50]   Bollinger L D and Böhmer H 1972 *Phys. Fluids* **15** 693

[5.51]   Arunasalam V, Heald M A and Sinnis J 1971 *Phys. Fluids* **14** 1194

[5.52]   Mase A and Tsukishima T 1975 *Phys. Fluids* **18** 464

[5.53]   Slusher R E and Surko S M 1980 *Phys. Fluids* **23** 472

[5.54]   Park H, Peebles W A, Mase A *et al* 1980 *Appl. Phys. Lett.* **37** 279

[5.55]   McIntosh G, Meyer J and Yazhau Z 1986 *Phys. Fluids* **29** 3451

[5.56]   Crowley T and Mazzucato E 1985 *Nucl. Fusion* **25** 507

[5.57]   TFR-Group and Truc A 1986 *Nucl. Fusion* **26** 1303

[5.58]   Nagatsu M, Nishida J, Ohnishi H, Tsukishima T, Okajima S,
         Mizuno K, Kawahata K, Tetsuka T and Fujita J 1987 *Nuclear
         Fusion* **27** 753

[5.59]   Dobrokhotov E I, Zharinov A V, Moskalev I N and Petrov D P
         1969 *Nucl. Fusion* **9** 143

[5.60]   Yelizarov L I and Zharinov A V 1962 *Nucl. Fusion* **2** Supplement
         669

[5.61]   Vlasov M A, Dobrokhotov E I and Zharinov A V 1966 *Nucl. Fusion*
         6 24

6 Experimental Data on Instabilities and Limiting Currents of Electron, Ion and Plasma Beams. Mechanisms of Current Limitation (Disruption) in Beams

6.1 The Limiting Electron Beam Current in the Absence of Space Charge Neutralization

The phenomenon of shut-off of the current of particles of identical sign by the space charge of this current has been known since the time of Bursian and Pavlov [2.2] and was often observed and reported by a number of researchers. The beam source (see figure 2.1) is a hot cathode C at a negative potential V_0 relative to an equipotential cylinder D with grounded walls and two grids at the end faces. The energy of electrons in the space between the grids is determined by the potential difference between the cathode and the left-hand grid of the cylinder D; it equals eV_0. The experiment is run in such a way that ions neutralizing the space charge of the beam cannot be produced during the time of observation. To achieve this, the accelerating potential difference V_0 is applied as a rectangular pulse whose length τ is much shorter than the characteristic time τ_i of the beam's space charge neutralization (see section 5.1):

$$\tau \ll \tau_i = (n_0 \sigma_i u)^{-1}. \tag{6.1}$$

To specify the geometry of the experiment, a strong magnetic field H of strength on the order of 10^3 Oe is applied along the axis of the cylinder D. The experiment consists of measuring the beam current while gradually increasing the cathode temperature. The beam current falls off abruptly

94

(i.e., is disrupted) after a certain temperature is reached, which indicates that the beam has been shut off by its own space charge: the so-called virtual cathode, formed in the beam, reflects a part of the beam back to the accelerating grid and the cathode. The beam current preceding the disruption is known as the *limiting current*. A result typical for such conditions was obtained in [6.1]. It states that the limiting current I_0 and the electron energy eV_0 obey a relation which coincides to within about 10% with the theoretical expression (2.11)

$$I_0 \simeq 32 \times 10^{-6}(eV_0)^{3/2}$$

where I_0 is given in amperes and eV_0 is in electron volts. This result holds for a geometry of experiment in which the radius of the cylinder D coincides with the beam radius. The experiment was carried out in the energy range of eV_0 from 2 to 500 eV.

Measurements of the limiting currents in another beam geometry, with the beam radius a being much smaller than the radius R of the cylinder D ($R/a = 1$ to 40), were performed in [6.2, 6.3] on an experimental set-up shown in figures 2.1 and 6.1. The results obtained are plotted in figure 6.2 for the energy range from 0 to 1200 eV, and in figure 6.3 for the energy range from 0 to 24 keV; in the latter case, the investigated energy range of the beam electrons is 50 times greater than in the experiments of [6.1]. The results shown in figures 6.2, 6.3 fit well the theoretical formula for a hollow beam [2.27] which is almost identical in this particular geometry to formula (2.10): the limiting current is

$$I_0 \simeq \frac{25 \times 10^{-6}(eV_0)^{3/2}}{\Delta/a + 2\ln(R/a)} \simeq \frac{25 \times 10^{-6}(eV_0)^{3/2}}{1 + 2\ln(R/a)} \quad \text{for} \quad \Delta \simeq a$$

where Δ is the thickness of the hollow beam, I_0 is in amperes and eV_0 is in electron volts.

The results are independent of the strength of the external longitudinal magnetic field in the range of 100–7000 Oe. In all the experiments described above, $L \gg R_0, a\dagger$.

An experimental analysis of the limiting currents at even higher (relativistic) energies of beam electrons ($eV_0 \geqslant 600$ keV) was carried out in [6.4–6.8]. The beam propagated in vacuum, there was no beam charge neutralization and a strong external magnetic field along the beam direction was present. The result obtained fits well the theoretical formula (2.30): the limiting current is

$$I_0 \simeq \frac{mc^3}{e} \frac{(\gamma_0^{2/3} - 1)^{3/2}}{1 + 2\ln(R/a)}$$

† The dashed curves in figures 6.2, 6.3 were plotted using a simplified formula, namely, the first term in the denominator was omitted (see [6.3]). The agreement of the experimental data with the theory becomes even better if this term is taken into account.

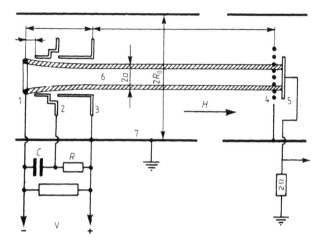

Figure 6.1 Schematic diagram of the experimental set-up for measuring the limiting currents in high-energy electron beams: 1–3, electron gun; 1, ring electrode; 2, intermediate ('pull-out') electrode; 3, accelerating electrode; 4, electron beam; 5, beam collector; 6, grid 7, vacuum chamber; $R = 200\,\text{Ohm}$, $C = 3\,\mu\text{F}$, V is the accelerating voltage. The magnetic field strength in the equipotential space is twice as high as at the cathode, so that the beam diameter (2.5 cm) is correspondingly smaller than the cathode diameter (3.5 cm).

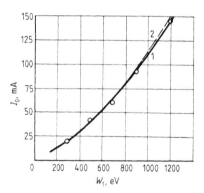

Figure 6.2 Limiting current I_0 in an electron beam free of neutralizing ions as a function of electron beam energy: $2R = 30\,\text{cm}$, $L = 100\,\text{cm}$, beam diameter $2a = 1\,\text{cm}$, $p = 1 \times 10^{-6}\,\text{mm Hg}$ [6.3]. The dashed curve plots the theoretical dependence given by (2.10).

where the relativistic factor $\gamma_0 = (1 - u^2/c^2)^{-1/2} = 1 + eV_0/mc^2$. If the beam current $I > I_0$, the beam is shut off by its own space charge. (On the effect of the magnetic field on the limiting beam current, see [6.9] and the literature cited therein.)

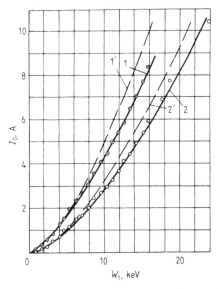

Figure 6.3   Limiting current I_0 in an electron beam free of neutralizing ions as a function of electron beam energy: 1, 1', $2a = 2.5\,\text{cm}$; 2, 2', $2a = 0.7\,\text{cm}$, $p = 1.4 \times 10^{-6}\,\text{mm Hg}$, $L = 120\,\text{cm}$, $2R = 30\,\text{cm}$, $H = 1500\,\text{Oe}$. Dashed curves plot formula (2.10). Current I_0 is independent of H [6.3].

The relationships outlined above that govern the changes in the beam limiting currents in the absence of space charge neutralization are practically identical to the theoretical predictions (see (2.8)–(2.11), (2.30)): the limiting beam current equals the Bursian current

$$I_0 = I_{\text{B}}$$

in the entire range of electron energies, up to relativistic values, if there is a sufficiently strong external magnetic field satisfying condition (2.32).

Quite numerous later observations of current shut-off in a relativistic electron beam by its own space charge after the current has reached the indicated value $I_0 = I_{\text{B}}$ were reported in the works devoted to developing the so-called vircators—extremely high-power generators of electromagnetic oscillations in the centimetre and millimetre wave bands. These devices of modern HF electronics which use the phenomena of formation and subsequent oscillations of a virtual cathode in a beam with over-Bursian current are discussed in chapter 11.

6.2   The Limiting Current of Quasineutral Electron Beam

Until the mid-1960s, the experimental data on the limiting currents of a quasineutral electron beam (with space charge compensated for by positive ions) were incomplete and largely contradictory (see also [2.6]). A systematic analysis of this problem had been undertaken in [2.8, 6.2, 6.3, 6.11, 5.28], in two stages. At stage I, an experimental set-up shown in figure 2.1 was used. A beam of (primary) electrons emitted by an indirectly heated tungsten cathode and accelerated in the electric field between two grids to an energy of several hundred electron volts ($W_1 = eV_0$), propagates along the axis of an equipotential (grounded) cylinder of stainless steel in the direction of an external time-independent magnetic field. The beam diameter is $2a = 1$ cm, the cylinder diameter is $2R = 30$ cm, the beam length L is controlled by moving an anode of diameter 15 cm in the interval from 5 to 150 cm. The equipotential volume in which the beam propagates is bounded at its ends by grounded grids with 2 mm square meshes on the side; these were made of tungsten wire 0.2 mm in diameter. The magnetic field distribution along the length of the beam is uniform (to within $\sim 3\%$), the field strength (H) is varied from 100 to 8000 Oe. The electron-beam space charge is neutralized by ions that the beam produces in the remaining gas whose pressure is about $(1-2) \times 10^{-6}$ mm Hg. To provide this degree of neutralization, three conditions need to be met. First, if the beam is in the pulse regime, the rate of increase of beam density must be below the rate of ion production, that is, a condition reversed with respect to (6.1) must be satisfied. Two beam regimes were initially used in the experiments of interest here: continuous and quasicontinuous (with current build-up time $\tau \simeq 2-3$ s); hence, the said condition was satisfied with a 'safety factor' of several orders of magnitude. To complete the picture, measurements of limiting currents were also carried out in these experiments in the pulsed regime of the beam, in the experimental arrangement of [6.11]. The time of (smooth) rise of current varied from several microseconds to several tens of milliseconds. Second, one has to eliminate the possibility of fast escape of ions from the beam to the end faces of the system along the external magnetic field. This condition was satisfied in the experiments [6.2, 6.3, 6.10] by using the so-called ion trap: a time-independent positive potential $V_c = (50-100)$ V was connected to the anode (the beam collector mounted behind the grounded grid) and to a specially introduced cathode grid (see figure 2.1). The third condition concerns the pressure of the neutral gas. The point is, that only those experiments can supply clear-cut data about the limiting currents of a quasineutral beam, in which the electron beam is not 'overcompensated', that is, the ion density does not exceed the beam density and thus slow (secondary) electrons are not present. In order to satisfy this condition, it is necessary to measure the limiting currents in the beam at a sufficiently low gas pressure, at which the beam has a small negative potential relative to the walls. This potential keeps the ions and repels the slow electrons (see (5.2)). This condition was also satisfied in the experiments we discuss.

The identification of the state of the beam in which a virtual cathode

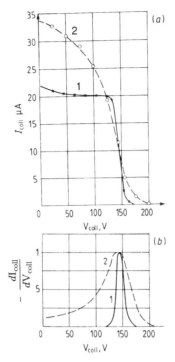

Figure 6.4 Examples of (*a*) retarding current–voltage characteristics and (*b*) beam-electron energy distributions. $W_1 = 150\,\text{eV}$. 1, beam current is lower than the excitation threshold for electron oscillations; 2, beam current is higher than this threshold. In case 1, the electron energy distribution is a delta-function [5.28].

is formed was carried out by the reversed current method described earlier in section 5.6.3. The presence of a grounded anode grid and a positive potential on the anode (the beam collector) removed, almost completely, any possible interference in beam current measurements by secondary electron emission from the anode.

The following methodological remark is needed before starting the description of how the limiting currents are measured in electron beams. High frequency (HF) electron oscillations which widen the electron beam velocity distribution and reduce the average velocity of beam propagation may be excited in the electron beam traversing a rarefied gas. Methods of eliminating these oscillations are also known: reduction of the pressure of the remaining gas, suppression of the secondary emission off the anode, shortening of the beam length. Having taken these precautions, the authors of [6.2, 6.3, 6.10] measured the limiting beam currents under conditions of almost complete suppression of HF oscillations, and the electron beam

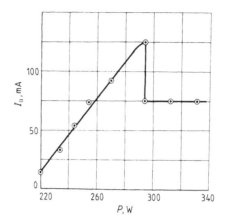

Figure 6.5   Beam current (at the anode) as a function of cathode heating power [6.2].

energy spectrum was very close to the δ-function (see figure 6.4).

The following procedure was chosen for measuring the limiting currents in the electron beam. The electron current at the beam collector was recorded at constant beam electron energy as a function of cathode temperature; the temperature was varied by varying the power of electron bombardment heating (figure 6.5). At a certain heating power, the electron current to the beam collector decreased jumpwise. The collector current immediately before the jump was assumed to equal I_{lim}, since the moment of jump coincided with the time of formation of the virtual cathode in the beam (this moment was established by the reversed current probe method). A typical oscilloscope trace recorded in a pulsed regime is shown in figure 6.6. Figure 6.7 shows the maximum beam current I_{max} as a function of the current rise time τ. We see that if $\tau > 300\,\mu$s, the maximum current is independent of τ. For this reason, the value of I_{max} at $\tau > 300\,\mu$s was assumed for the limiting current I_{lim}. Correspondingly, $I_{\text{max}} \simeq I_0$ at $\tau \simeq 5\,\mu$s.

Let us turn now to the experimental data obtained in [6.2]. Solid curves 2 and 3 in figure 6.8 show the limiting current, measured in a continuous regime, as a function of beam electron energy in different magnetic fields from 100 Oe (curve 2) to 6500 Oe (curve 3). The limiting currents at $100\,\text{Oe} \lesssim H \lesssim 6500$ Oe depend on electron energy in an absolutely identical manner and fall into the shaded area, between curves 2 and 3; H raised above 6500 Oe leaves the limiting current virtually unaffected (saturation). The theoretical curve 1 gives the limiting currents I_0 in a beam without ions; for the thin cylindrical beam in a large-diameter tube geometry $(L > 2R \gg 2a)$, relation (2.10) is valid. Curve 4 corresponds to the Pierce theory (see section 2.2) and to the expression (2.17) for the Pierce limiting

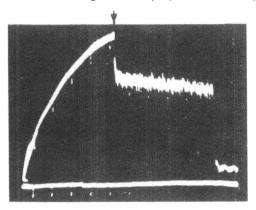

Figure 6.6 Oscilloscope trace of the (anode) beam current in the pulsed regime. The sweep time is 10 ms. The arrow marks the moment of current disruption. The abrupt current drop at the end of the oscilloscope is connected with the end of the current pulse [6.2].

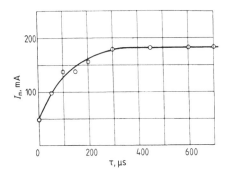

Figure 6.7 Maximum beam current as a function of current rise time τ. As $\tau \to 0$, $I_{max} \to I_0$, and at $\tau \gtrsim 300\,\mu s$, $I_{max} \to I_{lim}$ [6.2].

current I_P:

$$I_P \simeq \frac{130 \times 10^{-6}(eV_0)^{3/2}}{1 + 2\ln(R/a)}$$

where eV_0 is in electron volts and I_P is in amperes. The dashed curves 2 and 3 are obtained in the pulsed regime ($\tau = 600\,\mu s$) and correspond to the values of H of 100 and 6500 Oe, respectively. Dashed curve $1'$ was obtained in pulsed regime at $\tau = 5\,\mu s$. Clearly, in the range of not too small electron energies ($W_1 \geqslant 200$ eV), the limiting beam current I_{lim} is considerably smaller than the Pierce current I_P. The relationship of I_{lim} and I_P depends essentially on the magnetic field strength. For example, if

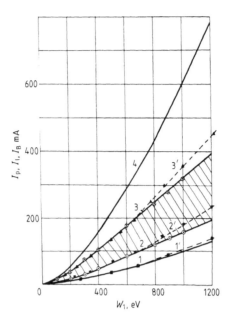

Figure 6.8   Limiting current as a function of electron beam energy in different magnetic fields [6.2, 2.8]; $2R = 30\,\text{cm}$, $L = 100\,\text{cm}$, $p = 1 \times 10^{-6}\,\text{mm Hg}$: 1, beam without ions, theoretical expression (2.10); $1'$, experiment in pulsed regime ($\tau = 5\,\mu\text{s}$, $H = 4000\,\text{Oe}$); 2 and 3, experiment in continuous regime, $H = 100$ and $6500\,\text{Oe}$, respectively; $2'$ and $3'$, experiment in pulsed regime ($\tau = 600\,\mu\text{s}$), $H = 100$ and $6500\,\text{Oe}$, respectively; 4, Pierce theory (2.18).

the magnetic field is low, I_{lim} is only a third or a fourth of I_{p} and differs only slightly from the current I_0 in the beam without ions. In strong magnetic fields (corresponding to the saturation of the curve $I_{\text{lim}}(H)$), the current I_{lim} exceeds I_0 by a factor of about 3.5 and approaches I_{p} but remains lower than I_{p} by a factor of about 1.5 to 2. In the range of low electron energies ($W_1 \lesssim 100\,\text{eV}$), the limiting currents in strong magnetic fields ($H \gtrsim 4000\,\text{Oe}$) coincide with the Pierce currents. Figure 6.8 shows that the limiting beam currents in the continuous regime coincide with those in the pulsed regime, and the value of I_0 measured in the pulsed regime at $\tau \simeq 5\,\mu\text{s}$ coincides with the theoretical value given by formula (2.10).

Figure 6.9 plots I_{lim} as a function of H at different values of beam electron energy. Owing to an appreciable dependence $I_{\text{lim}}(H)$, it is necessary to note that the function $I_{\text{max}}(\tau)$ shown in figure 6.7 is obtained at $H = 4000\,\text{Oe}$, which corresponds to the saturation of the curve $I_{\text{lim}}(H)$. Figures 6.8 and 6.9 show that the current I_{lim} exceeds I_0 not by a factor of 5.2 (as would be the case if the current restriction in the beam was

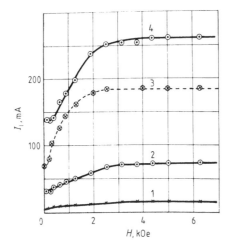

Figure 6.9 Limiting beam current as a function of magnetic field strength at different electron energies. $L = 100$ cm, $2R = 30$ cm, 1, 2, 4, continuous regime; 3, pulsed regime ($\tau = 300\,\mu s$) [6.2]. $W1 =:$ 1, 100 eV; 2, 300 eV; 3, 600 eV; 4, 900 eV.

determined by the Pierce instability (see section 2.2)) but only by a factor of at most 3.5.

Thus the experimental data shown in figures 6.8 and 6.9 show that in the entire range of variation of the parameters of the system (except in the range of low beam energies, $W_1 \lesssim 100\,\text{eV}$), the limiting beam currents are much lower than the Pierce currents given by formula (2.18).

A series of additional measurements of the limiting beam currents was carried out at two different tube diameters: $2R = 10$ and 6 cm. The results of these measurements are shown in figure 6.10. It demonstrates more clearly than figure 6.8 the changes in the dependence $I_{\text{lim}}(W_1)$, which occur both as W_1 increases and as H decreases.

The totality of the available experimental data thus allow us to summarize that the cases in which the limiting currents in electron beams are equal to the Pierce currents (i.e., are given by formula (2.18)) occur only in a relatively narrow range of experimental conditions and are exceptions rather than a general rule. If the parameters of a system vary in a considerably wider range, the restriction of the beam current sets in much earlier than the Pierce theory predicts. The very fact that there is this restriction (the disruption of the current) signifies that the neutralization of the beam space charge is violated even though all the necessary preliminary conditions have been satisfied. Since there were no factors in our experiments which could lead to a static decompensation of the space charge in the beam, we have to assume that the mechanism of

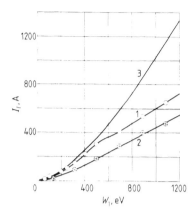

Figure 6.10   Limiting current as a function of beam electron energy for $2R = 6\,\mathrm{cm}$, $L = 10\,\mathrm{cm}$ [6.2]. 1, $H = 5200$ Oe; 2, $H = 1200$ Oe; 3, Pierce theory $I_1 = I_\mathrm{p}$.

violation of the beam neutralization is dynamic, that is, connected with the beam instability.

I will address this mechanism a little later (see section 6.3) and will now consider the results obtained at stage II of the experiment on measuring limiting currents in quasineutral electron beams [6.3]. A specific feature of these experiments, carried out on the same set-up, was a substantially higher electron energy (up to 20 keV) and varied beam diameter (from 0.7 to 2.5 cm). The electron gun (the beam source) operated in a single-pulse regime, with pulse length $\sim 2\,\mathrm{ms}$. The beam quasineutrality condition (5.1) (at a residual gas pressure $p \lesssim 1 \times 10^{-6}\,\mathrm{mm}$ Hg) was met quite satisfactorily: the beam had a negative potential of several volts relative to the surrounding walls and was thus well-neutralized by ions: the density of slow electrons (of gas origin) was negligible. For the beam to be neutralized by ions in the available time, the beam current was increased gradually: somewhat more slowly than linearly with the characteristic time $\tau \simeq 600\,\mu\mathrm{s}$; the beam neutralization condition (reversed with respect to (6.1)) was seen to be satisfied quite well. To generate the necessary shape of the current pulse of accelerated electrons in the beam, we chose an arrangement similar to that used earlier (see figure 6.1). The energy of electrons leaving the gun was determined by the accelerating voltage V_0 between the cathode and the third beam electrode. The V_0 pulse was rectangular, with $1\,\mu\mathrm{s}$ rise time. The smooth rate of beam current increase was determined by the shape of the 'pull-out' voltage pulse between the electron-gun cathode and the intermediate electrode 2: the length of the beam current rise time was dictated by the time constant RC for charging of the capacitor C which supplied the pull-out voltage.

Figure 6.11 gives typical oscilloscope traces of the beam current and the

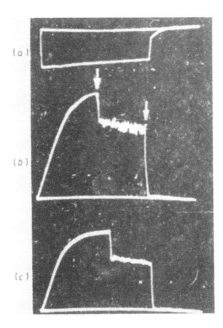

Figure 6.11 (*a*) Oscillograms of the accelerating voltage determining the beam electron energy and (*b*) and (*c*) beam current. The left-hand arrow points to the moment of current disruption and the right-hand arrow points to the moment when the beam pulse is terminated. The sweep time is 3 ms [6.3].

accelerating voltage which determined the electron energy.

In one series of experiments, the beam source was a planar tungsten cathode heated indirectly by electron bombardment; the cathode diameter was 1 cm. In other experiments, the source had a current-heated ring cathode made of tungsten or molybdenum wire 1.2 to 1.5 mm in diameter; the cathode diameter was varied in the range from 1 to 3.5 cm; in these experiments, the beam was hollow. The results obtained with both types of cathode of equal diameters coincided within experimental errors. The source was equipped with three electrodes (see figure 6.1). The second and third electrodes were either covered by wire grids as in [6.2, 6.10] (when the accelerating voltage did not exceed 3–4 kV) or had special holes for the passage of the beam [6.3]. The results obtained on sources containing grids and without them practically coincided. The beam electron energies W_1 corresponded to the accelerating voltage to within 5–10%; there was no interference to measurements from secondary electrons. The limiting current was not restricted by any processes in the beam source. This follows already from the experiment described below (see figure 6.15), which demonstrated that as the beam length was reduced from 100 to 10 cm, the

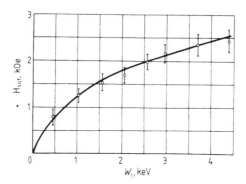

Figure 6.12   Magnetic field strength corresponding to approximate saturation on the curve $I_{\lim}(H)$ as a function of beam electron energy [6.3].

limiting current grew substantially (e.g., by a factor of two to three) and as the beam length was further reduced, the source current became insufficient for observing the current disruption (i.e., for generating an oscillogram of the type of figure 6.6). Therefore, current disruption is caused by processes in the equipotential space in which the beam propagates.

One of the characteristic results of experiments described above, with beams of relatively low energy ($W_1 \lesssim 1\,\mathrm{keV}$), was the fact that the limiting beam current $I_{\lim}$ essentially depended on the magnetic field strength. Namely, as H increases, the curve $I_{\lim}(H)$ tends to saturation (see figure 6.7). We can assume that this saturation is reached at a certain field strength $H = H_{sat}$ (e.g., in figure 6.7, $H_{sat} \simeq 3 \times 10^3\,\mathrm{Oe}$). H_{sat} as a function of electron energy W_1 was obtained in experiments at higher beam electron energies (figure 6.12). We see that the following relation fits the results quite well:

$$H_{sat} \propto (W_1)^{1/2}. \tag{6.2}$$

The dependence of the limiting current on electron energy in a wider range of parameter values than shown earlier is presented in figures 6.13, 6.14. In these figures, the parameters are the magnetic field strength and the diameter $2a$ of the hollow electron beam.

The effect of beam electron energy and beam diameter on the maximum current I_0 of non-compensated electron beam is shown in figures 6.2 and 6.3 (for obvious reasons, I_0 is independent of field strength).

The following conclusions can be drawn from figures 6.2–6.14.

1. The limiting beam current $I_{\lim}$ is greater than I_0 and less than the Pierce current $I_P \simeq 3\sqrt{3}I_0$. 2. The beam current $I_{\lim}$ at low electron energies varies approximately in proportion to $W_1^{3/2}$; at relatively high electron energies, $I_{\lim}(W_1)$ is an almost linear function of W_1: $I_{\lim} \propto W_1$. 3.

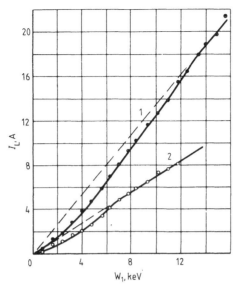

Figure 6.13 Limiting current as a function of beam electron energy. Nitrogen, $L = 120$ cm [6.3]. 1, $2a = 2.5$ cm; 2, $2a = 0.7$ cm. $H = 1500$ Oe.

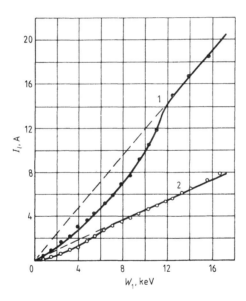

Figure 6.14 Limiting current as a function of beam electron energy. Nitrogen, $L = 120$ cm [6.3]. 1, $2a = 2.5$ cm; 2, $2a = 0.7$ cm. $H = 500$ Oe.

The $I_{\lim}(W_1)$ curve becomes linear at an energy which is greater, the larger the beam radius (and generally, the higher the magnetic field strength).

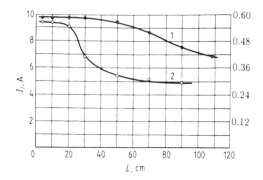

Figure 6.15   Limiting current as a function of beam length for various values of electron energy and magnetic field strength: 1, $H = 5000\,\text{Oe}$, $W_1 = 1\,\text{keV}$; 2, $H = 500\,\text{Oe}$, $W_1 = 6\,\text{keV}$. The left-hand coordinate axis refers to curve 2, the right-hand one, to curve 1 [6.3]. $2a = 2.5\,\text{cm}$.

4. As the beam radius increases, the limiting current I_{lim} grows much more than the maximal current I_0; in other words, the limiting current I_{lim} differs more from the current I_0, the greater the beam radius, the lower the electron energy W_1 and the higher the magnetic field strength.

5. The inflection on the $I_{\text{lim}}(W_1)$ curve is more pronounced, the lower the magnetic field strength. 6. The experimentally measured value of I_0 is quite close to the theoretical value calculated using (2.10) and some idealizing assumptions. This means that the maximum current in the beam without ions coincides with the limiting Bursian current ($I_0 = I_{\text{B}}$). 7. The limiting current I_{lim} decreases as H is reduced, while the maximum current I_0 is independent of H.

Figure 6.15 shows the typical dependence of the limiting current on the beam length L at various electron energies and magnetic field strengths. We find that as L increases, the limiting current decreases more, the lower H and the higher W_1. I have mentioned already that if the beam length is sufficiently small ($L \lesssim 10\,\text{cm}$), the current supplied by the electron gun becomes insufficient for shutting off (disrupting) the beam. Note that the greater the ratio W_1/H, the smaller the minimal beam length, beginning with which the current disruption is observed. Figure 6.15 clearly shows that the greatest current provided by the gun and measured for $L \leqslant 10$–$30\,\text{cm}$ varies in proportion to $W_1^{3/2}$. (The radial electron distribution corresponded to a hollow cylinder with wall thickness of 3 mm.)

I will repeat, to conclude this section, that for a long time, both before Pierce's theoretical paper [1.5] and after it appeared, it was believed that the restriction on the electron beam current (which sets in if there are no ions at $I = I_0$) can be removed by neutralizing the space charge of the beam by positive ions. Pierce was able to show theoretically, in contradiction to this opinion, that an instability must arise in the electron current whose space charge has been neutralized by ions already at a current $I = I_P = 3\sqrt{3}I_0$. A non-linear analysis of this instability [2.9, 2.11] led to the conclusion that the Pierce current I_P must be the limiting current of the neutralized electron beam. The experiments described here [6.2, 6.3, 6.10] demonstrated that the limiting beam current, whose space charge has been perfectly neutralized by ions, is even smaller than the Pierce current I_P; under certain conditions (high energy of beam electrons, moderately strong magnetic field and not too small beam length), this current is only slightly different from the maximum current I_0 in a beam without neutralizing ions. If, however, electron energies are sufficiently low and the beam fills the surrounding metal cylinder completely (as in [6.1]), the limiting beam current coincides with the Pierce instability threshold.

It is thus clear from the totality of the available experimental data that the limiting current in a quasineutral electron beam is restricted to a level which is certainly below the Pierce current, that is, it is restricted by an instability which is stronger than the Pierce instability. The nature and the mechanism of this instability are discussed in section 6.3.

6.3 Instabilities Responsible for Current Limitation (Disruption) in Quasineutral Electron Beams: Beam–Drift and Pierce Instabilities

An experimental analysis of the nature of instabilities that cause the current disruption in a quasineutral electron beam was carried out in [6.2, 6.3, 6.10]. The following two important factors were discovered relatively early in this study.

1. The instabilities in question are not connected with HF (electron–electron) oscillations since the beam disruption occurs also in situations when such HF oscillations are absent. Hence, when the beam disruption threshold is much lower than the Pierce current, the disruption is caused by a developing electron–ion instability.

2. The moment of shut-off of a quasineutral electron beam is identified very reliably by a reversed current probe (see sections 5.5.3 and 6.2). This means that an electric field with large azimuthal component E_φ arises in the process of beam shut-off. This implies that the instability responsible for the unexpectedly early current disruption in a quasineutral electron beam is connected with the generation of electron–ion oscillations that have no axial symmetry. Correspondingly, further study of the mechanism of current restriction in beams [2.8, 6.2, 6.3, 6.10] was aimed first of all at the experimental investigation of oscillations of this type, whose evolution would correlate with the restriction (disruption) of the current in the initially quasineutral electron beam. The spectra of these oscillations and

their amplitudes were measured by a series of disk probes 5 mm in diameter, which could be moved both along and transversely to the magnetic field (see figure 2.1) and were also used to find the concentrations of charged particles. Since the beam disruption occurs at a certain limiting current, the first to be studied were the conditions of occurrence (thresholds) of the 'suspicious' instabilities. The 'threshold' method of analysis proved to be very fruitful not only in analysing the causes of current restriction in beams; it will be shown below that this method made it possible to subject the theory of beam instabilities to a systematic experimental testing. I will illustrate this method with three examples: two electron–ion instabilities and one electron–electron. The experiments were run on the same set-up that was used to study the limiting currents in a quasineutral electron beam (see figure 2.1). The electron beam was run in a stationary regime.

The magnetic field strength was varied in the range from 100 to 8000 Oe and satisfied the condition

$$\omega_{Hi} \ll \omega \ll \omega_{He} \qquad k_z u$$

which signifies that electrons taking part in the oscillations (at a frequency ω) are magnetized while ions are not. The beam electron energy was varied in the range from 50 to 1000 eV, the (non-hollow) beam radius was 0.5 cm, the beam current was from several mA to several hundreds of mA, the gas pressure was 1×10^{-6} mm Hg. According to section 5.1, the beam in this situation is quasineutral ($n_+ \simeq n_1$) and the density of slow (secondary) electrons n_2 is very low in comparison with the beam electron density n_1. Some of the probes were used for a special study of the axial symmetry (or asymmetry) of oscillations. In these measurements, four probes were placed in the central cross section of the set-up at equal distances from the beam axis but at different azimuthal angles, 90° apart. An analyser of transverse ion energies (see figure 5.8) served as an indicator of participation of ions in these oscillations; to measure the ion and electron concentrations, the differential anode method was used (see section 5.6.1).

Experiments have demonstrated that there are no oscillations at frequencies $\omega \simeq k_z u$ under the conditions described ($p \lesssim 10^{-6}$ mm Hg, $n_2 \ll n_1$), and only low-frequency oscillations were found, at frequencies in the range $\omega_{Hi} < \omega < k_z u$, where $k_z \simeq \pi/L$; L is the beam length. These oscillations, whose typical spectrum is shown in figure 6.16, are characterized by different degrees of axial symmetry. The right-hand part of the spectrum in figure 6.16 represents oscillations with axial symmetry; these oscillations are treated below in section 6.4. The left-hand side of the spectrum represents oscillations without axial symmetry. The absence of this symmetry was very simple to demonstrate, namely, by non-coinciding signals of the probes at different azimuthal coordinates. Thus the signals of oppositely placed probes had opposite phases and the signals

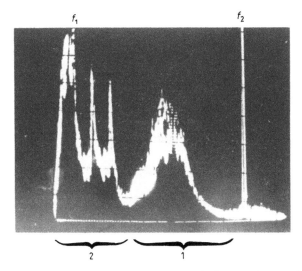

Figure 6.16   Frequency spectrum of electron–ion oscillations: 1, oscillations with axial symmetry; 2, oscillations without axial symmetry. Frequency reference marks: $f_1 = 0$, $f_2 = 1500\,\text{kHz}$, electron energy in the beam $W_1 = 150\,\text{eV}$, beam current $I = 25\,\text{mA}$, $H = 3500\,\text{Oe}$ [6.10].

of neighbouring probes (displaced azimuthally by a quarter of a full circle) differed by $90°$. The signs of these phase shifts corresponded to wave propagation in the direction of the Larmor rotation of the electron in the magnetic field.

These oscillations form standing waves along the beam, with the wavelength λ_z of the main harmonic equal to twice the beam length L, namely, $\lambda_z = 2\pi/k_z \simeq 2L$. This relation between λ_z and L is conserved as the beam length is varied†. The oscillations we discuss here are thus connected with the formation of a helical plasma 'tongue' which rotates in the 'electron' direction at a frequency equal to the observed frequency of oscillations. Typically, the azimuthal size of the tongue is 2 radians, and the amplitude of oscillations of the probe current is comparable with the DC component. Note that there is an excess of electrons at the advancing front of the tongue and an excess of ions at the back. Hence, there is an azimuthal charge separation and azimuthal electric field in the tongue. The strength E_φ of this field depends strongly on the beam current: it is several volts per centimetre at a relatively weak current but if the current exceeds the limiting value (when the beam shuts off), it reaches hundreds of V cm^{-1} (in fact, $E_\varphi \simeq W_1/ea$, where W_1 is the beam electron energy and a is the beam radius).

The oscillations in question are electron–ion ones: their generation is

† This situation is similar to that shown in figure 6.17.

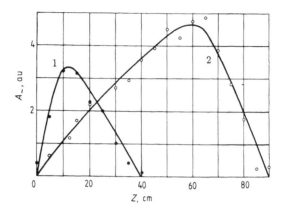

Figure 6.17   Distribution of the amplitude $A_\sim$ of electron–ion oscillations along the beam length; the probe coordinate z is measured relative to the beam source, 1, $L = 45\,\mathrm{cm}$, $f_s = 1.6\,\mathrm{MHz}$; 2, $L = 90\,\mathrm{cm}$, $f_s = 0.85\,\mathrm{MHz}$ [6.10].

accompanied with the acceleration of ions (perpendicularly to the direction of the beam velocity) to relatively high energies, of the order of the beam electron energy [6.12, 5.29].

The frequencies of these oscillations are quite close to the Langmuir ion frequencies: the characteristic frequency corresponding to the maximum amplitude of oscillations is $\omega \simeq \omega_+ = (4\pi n_+ e^2 / M)^{1/2}$, where M is the mass of a nitrogen atom (residual gas atom) and the ion density n_+ is equal to n_1, the electron density in the beam. As the beam current I varies, the oscillation frequency changes almost proportionally to $\sqrt{I}$. If the beam radius a is changed at a constant beam current ($I = $ constant), the oscillation frequency varies in inverse proportion to a: $\omega \propto 1/a$.

Let us now look at the most interesting aspect, namely, the relation of the oscillation amplitude to the beam current. The corresponding experimental curve is shown in figure 6.18. It is clear that oscillations generally survive at a sufficiently small beam current.

However, the oscillation amplitude starts to grow steeply at a sufficiently high beam current ($\simeq 46\,\mathrm{mA}$ in figure 6.18), and the beam current is soon disrupted: the reverse current probe described earlier indicates that a virtual cathode has been formed in the beam. The limiting beam current I_{lim} is 10–15% greater than the critical current I_c corresponding to a sharp growth in oscillation amplitude (in figure 6.18, $I_{\mathrm{lim}} \simeq 52\,\mathrm{mA}$). The relation between the currents I_{lim} and I_c is plotted in figure 6.19. The oscillations studied here are thus precisely the oscillations that cause the jumpwise disruption in the quasineutral electron beam. To be able to compare the experimental data with the theory later, it is also necessary to see how the current I_c depends on magnetic field strength (figure 6.20). We see that as H increases, the current I_c first rises steeply but then nearly reaches

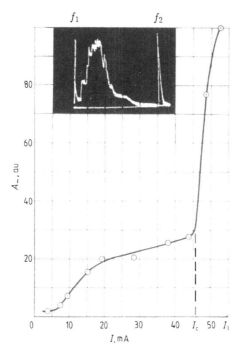

Figure 6.18 Amplitude of electron–ion oscillations as a function of beam current. $W_1 = 300\,\text{eV}$, $H = 2000\,\text{Oe}$, $L = 100\,\text{cm}$, $p = 2 \times 10^{-6}$ mm Hg. Shown in the inset above is the frequency spectrum of oscillations immediately before the beam current disruption ($I = 48\,\text{mA}$). The left- and right-most peaks are frequency reference marks: $f_1 = 0$, $f_2 = 1200\,\text{kHz}$ [6.10].

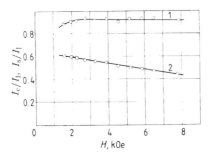

Figure 6.19 Ratios of the excitation thresholds of the beam-drift instability, I_c, and the Budker–Buneman instability, I_s, to the limiting beam current I_{lim}, as functions of magnetic field strength. $W_1 = 150\,\text{eV}$ [6.10, 2.8]. 1, I_c/I_1; 2, I_s/I_1.

saturation value which is practically equal to the theoretical value (2.18) of the Pierce current.

Let us move now to the interpretation, described above, of the

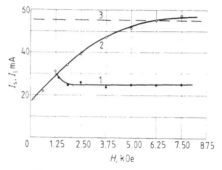

Figure 6.20   Excitation thresholds of electron–ion oscillations as functions of magnetic field strength, $W_1 = 200\,\text{eV}$, $L = 90\,\text{cm}$. 1, $I = I_s$—axially symmetric oscillations; 2, $I = I_c$—axially non-symmetric oscillations; 3, $I = I_P$—the threshold current for the Pierce instability [6.10, 2.8].

experimental data on those instabilities which cause the restriction (disruption) of the current in quasineutral electron beams [2.8, 6.10]. It is clear from the preceding description that to find out the nature of these instabilities, it is necessary first of all to compare the relations governing the changes in the limiting currents of beams and in the critical currents of excitation of axially non-symmetrical electron–ion oscillations, with the theoretical predictions of the thresholds of the Pierce and beam–drift instabilities. According to (2.18) and (2.19), the Pierce current of a beam shaped into a continuous rod of diameter a in a tube of radius $R \gg a$ and length $L \gg R$ is

$$I_P = \frac{ma^2}{4e} k^2 u^3$$

where, according to (2.14)),

$$k^2 \simeq k_r^2 = \frac{2}{a^2 \ln(R/a)}.$$

For the threshold (critical current) of the beam–drift instability, the theory gives (see (3.42))

$$I_c = \frac{ma^2}{4e} \frac{k^2 u^3}{1 + 2su/(a^2 \omega_{H_e} k_z)} = \frac{2I_P}{1 + 2su/(a^2 \omega_{H_e} k_z)}$$

where

$$k_z \simeq \pi/L \qquad k^2 \simeq k_r^2 + k_\varphi^2 \simeq 2k_r^2 \qquad S \simeq 1.$$

For a hollow cylindrical beam of radius a and thickness δ, used in experiments at high electron energy (up to 20–30 keV), expression (3.42) has to be somewhat modified:

$$I_c \simeq \frac{2I_P}{1 + CuL(\pi a\delta\omega_{H_e})^{-1}} \qquad (6.3)$$

where C is a numerical coefficient of the order of unity (for the first azimuthal oscillation mode with $S = 1$). The theory predicts the following dispersion properties of oscillations that are pumped as a result of the beam–drift instability. The range of oscillation frequencies is determined by formula (3.39). Oscillations do not possess axial symmetry; these are helical waves that propagate azimuthally in the direction of the Larmor rotation of electrons in the magnetic field; the wave number k_z is determined by the beam length and equals $k_z \simeq \pi/L$ for the main harmonic. Before carrying out the desired comparison of the experiments considered and the theory, I will summarize the observations and formulate the main conclusions on the changes in the limiting beam current in response to changing parameters and propagation conditions (see figures 6.6–6.20).

1. A range of conditions can be indicated in which the limiting current of the quasineutral electron beam equals the Pierce current or is quite close to it. These conditions are as follows: a sufficiently strong field H, a sufficiently small beam electron energy W_1, short beam length L, a relatively large beam radius a and a small gap between the beam and the surrounding metallic housing ($R \simeq a$). This conclusion holds under the conditions of figures 6.8 and 6.20 (at $H \geqslant 6.5 \times 10^3$ Oe, $L \leqslant 100$ cm and $W_1 \leqslant 100$ eV) and of figure 6.10 (at $H \geqslant 5 \times 10^3$ Oe, $L \leqslant 10$ cm and $W_1 \leqslant 500$ eV), and also under the conditions of [6.1] (in a sufficiently strong magnetic field). The limiting current displays the following properties in the outlined range of conditions:

$$I_{\lim} \propto u^3 \propto W_1^{3/2} \tag{6.4}$$
$$I_{\lim} \neq f(H) \tag{6.5}$$
$$I_{\lim} \text{ is independent of } L \text{ (if } L > R) \text{ and}$$
$$\text{is almost independent of } a. \tag{6.6}$$

2. Outside the indicated range of conditions, that is, in relatively weak magnetic fields (e.g., $H = (1000 - 1500)$ Oe), at sufficiently high W_1, small a, large L and a large gap between the beam and the surrounding housing ($R \gg a$), the limiting current behaves in a very different manner:

$$(a) \qquad I_{\lim} \propto H \tag{6.7}$$
$$(b) \qquad I_{\lim} \propto u^2 \propto W_1 \tag{6.8}$$
$$(c) \qquad I_{\lim} \text{ decreases greatly as } a \text{ decreases} \tag{6.9}$$

(see figures 6.13, 6.14), and is reduced much more than the maximum current of a non-neutralized beam (figures 6.2, 6.3);

$$(d) \qquad I_{\lim} \text{ decreases greatly as } L \text{ increases} \tag{6.10}$$

(e) $I_{\lim}$ is several times less than the Pierce current and can be close to the Bursian threshold ($I_0 = I_B$, see figure 6.8); $\qquad$ (6.11) (*f*) $I_{\lim}$ at sufficiently high H reaches saturation (approaching I_P), with the field strength H_{sat} that corresponds to this saturation changing as

$$H_{\mathrm{sat}} \propto u \propto \sqrt{W_1} \qquad (6.12)$$

(*g*) the electron energy W_1 at which the function $I_{\lim}(W_1)$ becomes linear and decreases as H and a decrease;

(*h*) the minimal beam length $L_{\min}$, beginning with which the beam disruption sets in, decreases as the ratio W_1/H is increased (if $L \ll L_{\min}$, $I_{\lim} \simeq I_P \propto W_1^{3/2}$).

It should be added that the maximum current I_0 in a beam without ions is independent of H (see figures 6.2, 6.3); hence, the ratio $I_{\lim}/I_0$ grows as H and a increase and as W_1 and L decrease, tending to the theoretical limit, that is, $I_P/I_0 = 3\sqrt{3}$ (see relation (2.17)). Finally, we take into account that oscillations that are sharply enhanced immediately before the beam current disruption are characterized by a helical structure with wave numbers $k_z \simeq \pi/L$ and $k_\varphi \simeq 1/a$; they are waves that rotate in the direction of the azimuthal rotations of electrons, at a frequency satisfying condition (3.39). The beam disruption is accompanied by the generation of electric fields with a very large azimuthal component $E_\varphi \simeq W_1/ea$. It is not difficult now to ascertain that the entire set of the properties of the limiting beam current listed above is in very good qualitative and quantitative agreement with the predictions of the theory about the excitation threshold I_c for the beam-drift instability. As we see from expression (3.42) for the critical (threshold) current I_c of excitation of the beam–drift instability, the dependence of this threshold on the parameters of the system is determined by the ratio of the terms in the denominator of (3.42). For example, if by virtue of (3.43) we have

$$2Su/a^2\omega_{H_e}k_z \gg 1$$

then

$$I_c \propto a^2 H W_1/L \qquad I_c \ll I_P \qquad (6.13)$$

and if

$$2Su/a^2\omega_{H_e}k_z \ll 1$$

then

$$I_c \simeq I_P \propto W_1^{3/2} \qquad I_c \neq f(a, H, L). \qquad (6.14)$$

It is rather easy to check that asymptotic relations (6.13) and (6.14) explain in an exhaustive manner the totality of observations, summarized above, of the dependence of the limiting beam current on electron energy, magnetic field strength, beam radius and beam length, etc. According to the approach outlined here, the effect of the gap $(R - a)$ on the limiting current is explained by the fact that if $R \simeq a$ (the beam slides along metallic walls), the axially asymmetric oscillations of the fundamental mode $(S = 1)$ cannot be excited since we have $E_\varphi = k_\varphi = 0$ on the beam surface.

It is thus clear that the restriction (disruption) of current in a quasineutral electron beam is caused by the beam–drift instability when the limiting current I_{lim} is considerably less than the Pierce threshold I_{P}, and by the Pierce instability if $I_{\text{lim}} \simeq I_{\text{P}}$. This conclusion also describes quite well the facts that are connected with the generation of axially symmetric electron–ion oscillations for which it will be possible to show (see section 6.4) that they are generated owing to the Budker–Buneman instability whose threshold is I_{s} (see (3.14)). As follows from (3.14) and (3.42), under typical conditions of these experiments and at high field H, the current I_{c} must be approximately twice as large as I_{s}. Figure 6.20 shows that the experiment points to just this ratio of the thresholds I_{c} and I_{s}.

To make the picture complete, we need to bear in mind the following factor. The theory of excitation of the beam drift instability as presented above is a linear one. It predicts the instability threshold I_{c} (see formula (3.42)) but says nothing about its macroscopic consequences, especially about such a highly non-linear phenomenon as the disruption of the beam current. We have seen, however, that experiments demonstrate that if the beam current exceeds the maximum current I_0 in the beam without ions, the threshold I_{c} given by theoretical formula (3.42) is almost equal to the limiting beam current (see figure 6.19) in the entire range of beam electron energies investigated (i.e., at least up to 20 keV). As for the Pierce instability, its non-linear theory [2.9, 2.11] shows that the Pierce threshold I_{P} must be the limiting beam current: if $I > I_{\text{P}}$, a virtual cathode is formed in the beam (of course, if the parameters of the system are such that the virtual cathode does not develop earlier, as a result of the evolution of the beam–drift instability). It is possible to conclude, therefore, that almost complete clarity has been achieved in understanding the nature of the instabilities responsible for the current restriction (disruption) in quasineutral electron beams†.

It should be pointed out specially that the beam–drift instability (studied for the first time in the experiments described above) is a rather 'crude' phenomenon, not very sensitive to subtleties of the experimental situation. For instance, we will be able to see later (see section 6.9) that it arises not only in monoenergetic beams but also in beams with high velocity spread, and even in beams with a uniform velocity distribution of electrons (of the type of a symmetric plateau) in the interval $\pm u_0$, where u_0 is the maximum electron velocity.

I will emphasize the following to conclude this section. Since the beam–drift instability of a quasineutral beam and the drift instability of a

† See section 6.8.

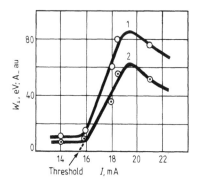

Figure 6.21   Ion energies $W_\perp$ (2) and amplitudes $A_\sim$ of axially symmetric electron–ion oscillations (1) as functions of beam current I. The ion energy analyser is located at the 8 cm radius, $H = 2600$ Oe, $L = 90$ cm, $p = 1 \times 10^{-6}$ mm Hg, $W_1 = 150$ eV; gas: nitrogen [6.10].

Maxwellian plasma are based on essentially the same common mechanism (see section 3.4), the experiments discussed above can be regarded as modelling the (universal) drift instability of non-uniform plasma in the favourable case of a beam, that is, of a δ-shaped electron energy distribution function. This modelling is of principal importance since the drift instability plays a fundamental role in solving problems of controlled nuclear fusion.

6.4   Budker–Buneman Electron–Ion Instability. Does it Produce Current Disruption?

I will now describe the results of experimental analysis of the electron–ion instability [6.2, 6.10, 2.8].

Section 6.3 treated that part of the spectrum of electron–ion oscillations which is connected with the axially non-symmetric modes. These modes correspond to the left-hand side of the common oscillation spectrogram shown in figure 6.16. Let us turn now to the right-hand side of this spectrogram. The same method that was applied to analysing the spatial structure of axially non-symmetric oscillations, revealed that the oscillations under consideration were axially symmetric: the phases of signals of all four probes separated by the azimuthal angle of 90° are identical. These are electron–ion oscillations: when they are produced, they accelerate ions to energies of the order of the beam-electron energy, just as the axially non-symmetric oscillations do; the behaviour observed in this case is, in principle, similar to that shown in figure 6.21.

The dispersion characteristics of axially symmetric oscillations are shown in figures 6.22–6.25. Figure 6.22 gives frequency spectra of oscillations, recorded at three values of beam current. We see here (see also figure 6.23) that the oscillation frequency decreases as the current increases. Figures 6.24, 6.25 plot oscillation frequency as a function of electron velocity u and beam length L. In a large part of the range of variation of these parameters, a qualitative regularity is observed which can be approximated roughly by the relation

$$f_s \propto u/L. \tag{6.15}$$

The oscillation frequency f_s depends very weakly on the magnetic field strength. Just as the axially non-symmetric oscillations, these oscillations form standing waves along the beam (see figure 6.17). The wavelength of the fundamental (first) harmonic is approximately equal to twice the beam length: $\lambda_z \simeq 2L$, $k_z \simeq \pi/L$. By virtue of relation (6.15) and the data of figures 6.22–6.25, we obtain

$$\omega_s = 2\pi f_c \simeq \left(\tfrac{1}{4} - \tfrac{1}{5}\right) k_z u. \tag{6.16}$$

A feature of principal importance is that the excitation of axially symmetric oscillations starts at a certain threshold current I_s. Figures 6.26, 6.20 show the threshold current I_s as a function of electron velocity and magnetic field strength. The following clearly pronounced facts attract one's attention:

$$I_s \propto u^3 \qquad I_s \neq f(H) \qquad I_s \simeq \tfrac{1}{2} I_P \tag{6.17}$$

where I_P is the theoretical value of the Pierce limiting current (see (2.17)).

Now it is not difficult to use the experimental results presented to ascertain that all the described properties of oscillations (excitation thresholds and their dependence on system parameters, the spatial structure and the dispersion characteristics as a whole) are indeed in good agreement with the properties of Budker–Buneman electron–ion oscillations whose theory was presented in section 3.2. This can be seen from a comparison of equations (6.15)–(6.17) with the theoretical relations (3.25) and (3.27) where the following experimental data must be used: $k_z \simeq \pi/L$, $L \simeq 100\,\mathrm{cm}$, beam radius $a \simeq 0.5\,\mathrm{cm}$, $R = 15\,\mathrm{cm}$, $k^2 \simeq k_r^2 \simeq 2.4\,\mathrm{cm}^{-2}$, $M/m = 5.3 \times 10^4$ (nitrogen) and

$$\left(\frac{M}{m} \frac{k_z^2}{k^2}\right)^{1/3} \simeq 2.5. \tag{6.18}$$

Note, among other things, that in agreement with the theory of the Buneman instability, the frequency of the oscillations in question decreases as the beam current increases (see figure 6.23); the ratio of the oscillation frequency at a given beam current ($I > I_s$) to the frequency ω_s of oscillations at the instability threshold ($I = I_s$) fits quite well the

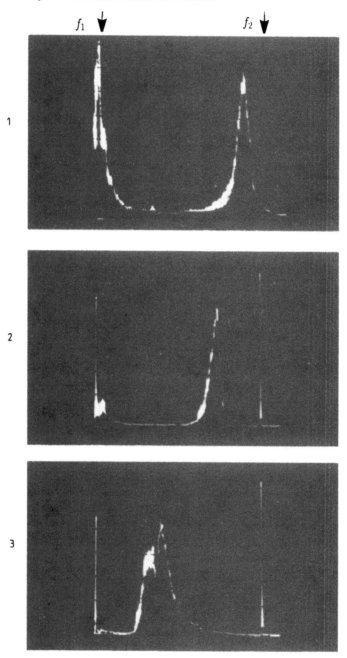

Figure 6.22   Frequency spectra of axially symmetric oscillations. Frequency marks: $f_1 = 0$, $f_2 = 1500\,\text{kHz}$, $W_1 = 300\,\text{eV}$,
$H = 4000\,\text{Oe}$: 1, $I = 52\,\text{mA}$; 2, $I = 59\,\text{mA}$; 3, $I = 64\,\text{mA}$. The
oscillation amplitudes in the 1, 2, 3, cases are in ratio 1:2.5:2.5.
The threshold current was $I_s = 50$ mA [6.10].

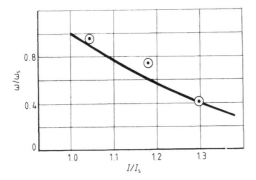

Figure 6.23 Ratio of the frequency ω of Buneman oscillations at a beam current I greater than the threshold I_s to the frequency ω_s of oscillations at the oscillation threshold (at $I = I_s$), as a function of the ratio I/I_s [6.10].

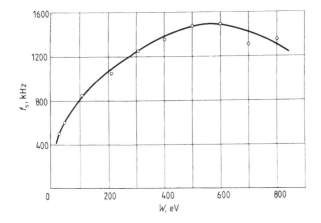

Figure 6.24 Frequency f_s of axially symmetric electron–ion oscillations as a function of beam electron energy. The oscillation frequency is measured at the critical beam current ($I = I_s$). $L = 90$ cm [6.10].

theoretical curve of figure 6.23 reproduced from figure 3.8. At the same time, one immediately identifies a fact which has already been emphasized in the introduction: the measured values of instability thresholds agree much better with the predictions of the theory than is found for the experimental data on the dispersion characteristics of the electron–ion oscillations. This happens because the dispersion properties are measured in the regime in which the oscillations have a finite (not too small) amplitude and are thus sensitive to non-linear effects, while instability thresholds are obviously independent of these effects. This result is an additional evidence of the advantages of the threshold method of experimental study of beam

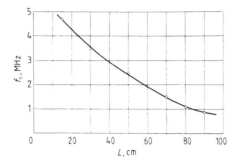

Figure 6.25   Frequency f_s of axially symmetric electron–ion oscillations as a function of beam length L. The oscillation frequency is measured at the critical beam current $(I = I_s)$. $W_1 = 150$ eV [6.10].

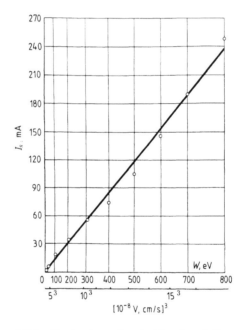

Figure 6.26   Excitation threshold of axially symmetric electron–ion oscillations as a function of velocity cubed of beam electrons: $H = 7800$ Oe, $L = 90$ cm, $p = 1.4 \times 10^{-6}$ mm Hg [6.10, 2.8].

instabilities.

The following fact of principal importance deserves special attention: in contrast to the axially non-symmetric oscillations discussed in the preceding section, those of this section do not lead to a restriction (disruption) of the beam current. This fact signifies that the Budker–Buneman instability does not result in the formation of a virtual cathode in the beam.

The experiments [6.2, 6.10] are the first in which the electron–ion Budker–Buneman instability has been unambiguously identified.

The experimental data discussed in the last two sections show that the 'threshold' method developed in [2.8, 6.2, 6.10] makes it possible to separate reliably various instabilities inherent to the quasineutral beam and to study them in their 'pure' form, systematically and unambiguously. The following factor is largely responsible for these advantages [6.12]. If the magnetic field is sufficiently strong (it magnetizes electrons but not ions), the small additional term $(m/M)^{1/3}$ (occurring in the case of the planar geometry, see (3.14)) is replaced in the denominator of expression (3.27) for the Budker–Buneman instability threshold by a considerably greater term $(\frac{m}{M}k^2/k_z^2)^{1/3}$, as a result of which the Budker–Buneman instability threshold becomes lower by a factor of $[1 + ((m/M)k^2/k_z^2)^{1/3}]^3$ that is, by about 2–2.5 times) than the Pierce instability threshold: the instabilities do separate! This behaviour must be regarded as extremely favourable: without it (i.e., in the absence of the longitudinal magnetic field), the two instabilities would superpose, which would greatly complicate their separate analysis. Furthermore, it has already been mentioned that the Pierce instability threshold is clearly defined only in the presence of a strong longitudinal magnetic field: without such a field, the threshold is substantially increased by the strong divergence of electron trajectories (owing to their mutual repulsion) [2.28, 2.29].

The experimental result discussed above is of principal importance: it indicates that the Budker–Buneman instability in cylindrical geometry does not result in beam current disruption. This result may seem to contradict the prediction of the theory outlined in the first part of section 3.2. However, as mentioned in the second part of section 3.2, the indicated conclusion of the theory on the possibility of current disruption due to the Buneman instability is valid only for the resonant regime $(k_z \gg k_\perp)$ while in the experimental conditions discussed here we have, on the contrary, $k_\perp \gg k_z$. A possible interpretation of the absence of disruption in cylindrical geometry appears to be as follows. The beam current disruption may occur owing to the capture of its electrons by a wave developing under instability and having a very low phase velocity, much lower than the unperturbed velocity of beam electrons. It is natural to assume that the capture of electrons by a wave occurs on the interaction length *of at least several wavelengths*. In fact, the beam length in the experiments we discuss here is only one half of the wavelength—see figure 6.17. Therefore, it seems quite likely that the interaction length of beam electrons and the wave is too short for the capture and disruption of the beam current.

Another interpretation of the absence of disruption of the total beam current in the Budker–Buneman instability has been given in a theoretical paper [6.13], according to which the current disruption does take place but not in the entire beam: only close to the axis. Additional experimental analysis is very welcome in this case.

An experimental observation of the Budker–Buneman instability in toroidal plasma was also reported in [6.14]. The electron beam was generated by the electric field produced by an alternating magnetic field. The experiment showed that as the beam current reached the threshold of the instability in question, the restriction (disruption) of growth of the current took place. This effect occurred if the flow velocity of electrons was greater than their thermal velocity. After the disruption of the current, beam electrons get heated and their random velocity becomes roughly equal to the flow velocity; the system then reaches the stability limit. Under the specific conditions of this experiment (electron energy $W_1 \simeq 250\,\mathrm{eV}$, beam current $I \simeq 120 - 150\,\mathrm{A}$), $k_\perp \simeq 1/3\,\mathrm{cm}^{-1}$ and $k_z \simeq \omega_1/u \simeq 15\,\mathrm{cm}^{-1}$, that is, $k_z \gg k_\perp$, that we recognize as the case of the *resonant instability regime*†. The geometry of this experiment is, to very high accuracy, one-dimensional. The conditions chosen in this experiment and the reported result are in agreement with the predictions of the theory. At the same time, these conditions differ drastically from the experimental conditions in cylindrical geometry, which, as has been mentioned already seemed to have been the cause of the differences.

6.5   The Electron–Electron Instability

The advantages of the threshold method of experimental investigation of beam instabilities were outlined in the preceding sections. This method was also applied in [6.15] to studying the generation of electron–electron oscillations. The plasma density (formed by the beam in the gas) was less than, or of the order of, the beam density, so that the spectra of oscillations generated by the beam (at currents above the threshold) were found to be very different from those of the familiar Langmuir oscillations in plasma; namely, they were determined to a much higher degree by the properties of the beam itself. The experiments were conducted in the same experimental set-up that was used for measuring the thresholds of the electron–ion and beam–drift instabilities and the limiting beam currents (see figure 2.1).

In addition to the threshold current of the instability, the spectra of electron oscillations were also recorded in these experiments. Oscillations of the electron current to a needle (Langmuir) probe were used as indicators in these measurements. The probe was made of molybdenum wire 4 cm long and 0.8 mm in diameter; it intersected the beam along its diameter and could be shifted along the beam axis. The oscillation spectra of the total

† There was not enough time during the observation period for the secondary plasma to form, so the system was very nearly a quasineutral electron beam.

electron current to the beam collector had the same shape as the probe current spectra.

The analyser screen recorded an oscillogram of the spectral function in the amplitude–frequency coordinates, with linear scales on both axes. Figure 6.27 shows typical spectra of HF oscillations at a relatively low current ($I = 4 - 7\,\text{mA}$) and low energy of beam electrons ($W_1 = 100 - 200\,\text{eV}$). These spectra display the following clearly pronounced common features:

1. Each spectrum has pronounced maxima at discrete frequencies f_n ($n = 1, 2, 3, \ldots$).

2. All intervals between the discrete frequencies, Δf_n, are the same, to within 10–25%, and equal the first frequency f_1:

$$\Delta f_n = f_n - f_{n-1} = f_1. \qquad (6.19)$$

In other words, all frequencies f_n are, to within the accuracy indicated, the harmonics of the frequency f_1. These common features of the oscillation spectrum remain unchanged at all beam parameters, provided the current I is below the limit I_{lim} at which a virtual cathode is formed in the beam. The condition $I < I_{\text{lim}}$ was satisfied in all cases described here. In a number of experimental runs, the oscillation spectrum was studied as a function of beam parameters. It was found that the relation $f_n = n\Delta f$ invariably held, so that measuring Δf is sufficient for characterizing the frequency spectrum.

Figure 6.28 plots Δf as a function of beam electron velocity for two values of the beam length. We see that in both cases, Δf is proportional to u, and that the proportionality coefficient changes in the transition from $L = 80\,\text{cm}$ to $L = 150\,\text{cm}$ as $1/L$. Figure 6.29 plots Δf as a function of beam length for two values of velocity of primary electrons: (a) $u = 1.04 \times 10^9\,\text{cm}\,\text{s}^{-1}$ and (b) $u = 0.73 \times 10^9\,\text{cm}\,\text{s}^{-1}$. Clearly, Δf is proportional in both cases to the inverse beam length $1/L$, and the proportionality coefficient changes in the $a \rightarrow b$ transition proportionally to u. Therefore, figures 6.28 and 6.29 and relation (6.19) imply that

$$\Delta f = c_1 u / L \qquad f_n = c_1 u n / L \qquad (6.20)$$

where c_1 is a coefficient dependent on the beam and plasma densities and of the beam velocity spread.

A number of experimental runs were devoted to studying the conditions leading to the generation of oscillations. The following conclusions were drawn. First, it was necessary that the beam length L exceed a certain 'critical' value L_c equal to 20–40 cm in our experiments. Beginning with $L = L_c$, an increase in L increases the oscillation amplitude (say, up to $L = 75\,\text{cm}$) after which the amplitude starts to fall off gradually.

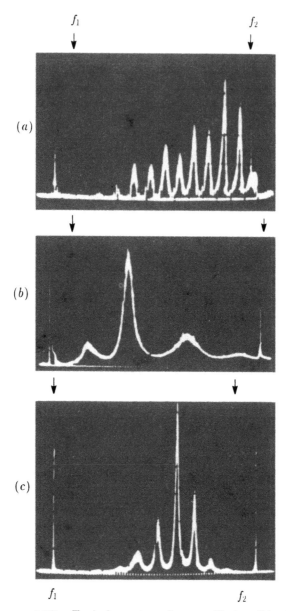

Figure 6.27   Typical spectra of HF oscillation (*b*) of the anode current and (*a*) and (*c*) of the needle probe current. Two narrow high lines on the left and on the right are frequency reference markers: $f_1 = 0$ and f_2 (shown by arrows). Low frequency oscillations (*b*) are seen at the f_1 marker: (*a*) $f_2 = 30$ MHz, $W_1 = 200$ eV, $I = 7$ mA, $L = 130$ cm; (*b*) $f_2 = 30$ MHz, $W_1 = 500$ eV, $I = 51$ mA, $L = 43$ cm; (*c*) $f_2 = 16$ MHz, $W_1 = 140$ eV, $L = 140$ cm, $p = 5 \times 10^{-6}$ mm Hg [6.15].

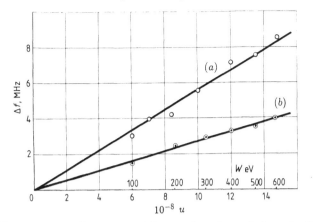

Figure 6.28 Frequency Δf as a function of the velocity of primary electrons for two values of beam length [6.15]. a, $L = 80$ cm: b, $L = 150$ cm.

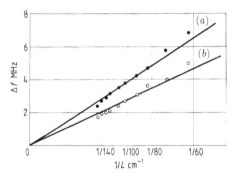

Figure 6.29 Frequency Δf as a function of the inverse beam length for two values of energy of primary electrons [6.15]. a, $u = 1.04 \times 10^9$ cm s^{-1}; b, $u = 0.74 \times 10^9$ cm s^{-1}.

Second, the beam current must exceed a certain threshold (a certain critical current) $I > I_c$, whose value is dictated by the parameters of the system (figure 6.30). The instability threshold depends most strongly on the velocity of beam electrons. Figure 6.31 shows that this dependence is given by the formula

$$I_c \propto u^3. \tag{6.21}$$

The same figure shows that as the gas pressure increases, the threshold decreases.

Figure 6.32 gives the amplitude of a fixed harmonic ($n = 5$) of current oscillations at the needle probe located on the beam axis, as a function of the distance between this probe and the cathode (beam length $L = 135$ cm). Clearly, the oscillation amplitude reveals a well-pronounced periodicity

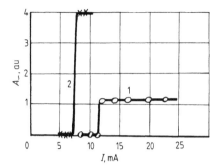

Figure 6.30   Oscillation amplitude $A_\sim$ of one of the harmonics ($f = 16\,\text{MHz}$) of the needle probe current as a function of beam current, for two values of the hydrogen pressure: 1, $p = 5 \times 10^{-6}\,\text{mm Hg}$; 2, $p = 3 \times 10^{-5}\,\text{mm Hg}$. $W_1 = 300\,\text{eV}$, $L = 135\,\text{cm}$ [6.15].

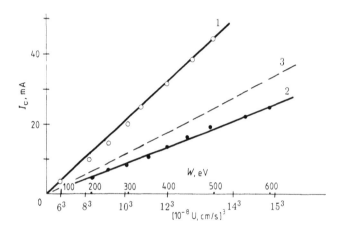

Figure 6.31   Threshold current I_c as a function of velocity cubed of primary electrons, for two values of the hydrogen pressure: 1, $p = 2.4 \times 10^{-6}\,\text{mm Hg}$; 2, $p = 3 \times 10^{-5}\,\text{mm}$ Hg; 3, theoretical dependence. W_1 is the beam electron energy, $L = 150\,\text{cm}$ [6.15].

along the beam. The spatial period (half of the wavelength, $\lambda_n/2$) equals 25–28 cm for $n = 5$, that is, it satisfies the relation

$$\lambda_n = 2L/n. \tag{6.22}$$

Figure 6.32 compares the variation in oscillation amplitude along the beam for two harmonics: (b) $n = 5$ and (c) $n = 8$. (This comparison has been carried out only for the second half of the beam length ($70\,\text{cm} \leqslant L \leqslant 135\,\text{cm}$), where the amplitudes of the two harmonics were sufficiently

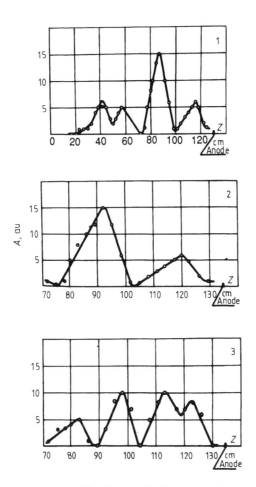

Figure 6.32 Amplitude of a fixed current harmonic on the needle probe as a function of probe–cathode distance: 1 and 2, $n = 5$, $f = 10\,\text{MHz}$; 3, $n = 8$, $f = 16\,\text{MHz}$, $W_1 = 300\,\text{eV}$, $L = 135\,\text{cm}$.

large.) Clearly, the spatial periods of oscillations of these two harmonics are 25–27 cm for $n = 5$ and 15–17 cm for $n = 8$, that is, are equal to $L/5$ and $L/8$, respectively. The data of figure 6.32 thus allow a conclusion that the wavelengths of different harmonics are related to the beam length by equation (6.22); this means that standing waves are formed in the system: the beam length equals an integral number of half-waves. The formation of standing waves is caused by the reflection of running waves from metal electrodes which define the beam length.

Let us move now to comparing the obtained experimental data with theoretical relations (3.32) and (3.33) for the frequency and excitation threshold of electron–electron oscillations. These relations include a quantity k_z determined from the experimental result (6.22):

$$\cdot k_z \equiv 2\pi/\lambda_z = n\pi/L \qquad n = 1, 2, 3, \ldots. \tag{6.23}$$

In view of (6.23), equation (6.22) yields a theoretically predicted spectrum of electron–electron oscillations,

$$\omega = \omega_n \simeq \pi \frac{u}{L} n \frac{1}{1 + \alpha^{-1/3}} \tag{6.24}$$

where $\alpha = n_2/n_1 \gg m/M$. According to (6.24), the oscillation frequencies ω_n are proportional to n, u and $1/L$. The experimental result (6.20) shows that experiments do reveal just such qualitative behaviour. As α increases (this is achieved by increasing the gas pressure), the slow growth of the oscillation frequency predicted by (6.24) is indeed observed in the experiment. As for the quantitative aspect of the problem, the observed oscillation frequencies (see figures 6.28 and 6.29) are found to be smaller by a factor of 1.2 to 1.7 than frequencies (6.24) predicted theoretically for a monoenergetic beam (uniform over the cross section): if the experimental data shown in figures 6.28 and 6.29 by circles, are multiplied by this coefficient, we obtain the theoretical result plotted in these two figures by straight lines. Hence, the experimental data on the dispersion characteristics of the oscillations under consideration are in good agreement with the theory, namely, to within a (constant) numerical coefficient of 1.2–1.7 (this value is easily explainable by the decrease in the average velocity of beam electrons due to the deceleration on the oscillations generated—see figure 6.4). This agreement is naturally much better than for the electron–ion oscillations (see section 6.4).

Let us look at the instability threshold. The theoretical dependence $I_c(u^3)$ calculated using (3.32) for $\alpha = 1$ is plotted as a dashed curve, 2, in figure 6.31. The same figure gives the experimental data for two values of gas pressure. Obviously, the theoretical dependence $I_c(u^3)$ is quite close to the experimental curves. If we assume that the value of α on curve 1 is somewhat less than unity and the value on curve 2 is somewhat greater than unity, we obtain not only a qualitative but also a quantitative agreement with the experimental data (as for the true value of α, I have already mentioned that it is not very different from unity). It is also necessary to point out that according to (3.32), the instability threshold is determined by the total wave number $k \simeq 1/a$ and thus should not appreciably depend on $\lambda = 2L/n$, that is, on the order of the spectral harmonic. This is also in agreement with the experiment. We also see from (3.32) and (3.27) that if $\alpha \gg m/M$ (e.g., if $\alpha \simeq 1$), the electron–electron instability threshold may drop considerably lower than the electron–ion instability threshold, which is in its turn twice lower than the Pierce threshold. Under the conditions of the experiments described, the threshold I_c was several times less than the Pierce current.

The experiments [6.15, 5.28] have demonstrated that the energy spectrum of beam electrons depends essentially on the beam current; if the beam current I is not higher than the electron–electron instability threshold I_c, the beam is almost monoenergetic; if $I > I_c$, the velocity distribution function of beam electrons is broadened (figure 6.34). If $I > I_c$, the beam velocity spread depends very strongly on the gas pressure: if $p \simeq 1 \times 10^{-6}$ mm Hg, then $\Delta u \simeq (1/4)u_{av}$ (where u_{av} is the average directed velocity of the beam), and if pressure is increased to $p \simeq 1 \times 10^{-4}$ mm Hg, then Δu becomes comparable to u_{av}.

It, has been mentioned already that the oscillations discussed here differ substantially from the Langmuir oscillations, both in the spectrum and in that they are excited only when the beam density n_1 is not negligible in comparison with the plasma density n_2. The frequencies of the oscillations observed are of the order of 10^7 Hz, that is, are lower by a factor of several tens than the electron Langmuir frequency $f_p = \omega_p/2\pi$. Experiments demonstrate that if the plasma density is substantially increased (by appropriately increasing the gas pressure), the maximum of the oscillation amplitude shifts to considerably higher frequencies (to shorter wavelengths). This fact is apparently an expression of the tendency to a gradual transition of the oscillations discussed above to the Langmuir oscillations. There are grounds for assuming that if n_2/n_1 is substantially increased, long-wavelength oscillations will vanish and only Langmuir oscillations will be excited in the system, as is the case of a tenuous beam propagating through a dense plasma.

I have already mentioned in section 3.3 that by virtue of a theoretical relation (3.31), there exists a threshold plasma density n_{2c}, such that if $n_2 > n_{2c}$, the electron–electron oscillations are excited at any beam current (from which only their increment depends). This prediction of the theory was confirmed in [6.16] (see also an earlier paper [6.17]). According to formula (3.31) of the theory, $n_{2c} \propto u^2$ when the magnetic field is sufficiently strong.

The contents of all five sections of this chapter thus show that the linear theory of electron–ion and electron–electron instabilities presented in chapters 1–3 is in good agreement with the experimental data on the thresholds of beam excitation of oscillations and on their dispersion properties. The reader shall recall that this theory included a simplification, connected with applying the quasiclassical approximation to large-scale perturbations ($k_r a \simeq 1$, $k_z L \simeq 1$). The comparison of this theory with the experimental data given above has demonstrated that this simplification is definitely justified, and that the error caused by it is negligible (for relevant details, see section 6.8).

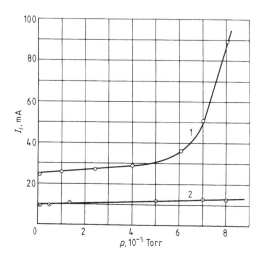

Figure 6.33   Limiting electron beam current as a function of
gas pressure. $W_1 = 150\,\mathrm{eV}$, $H = 1300\,\mathrm{Oe}$, $L = 10\,\mathrm{cm}$: 1, beam
radius 0.5 cm; 2, beam radius 0.25 cm. If $p = 8 \times 10^{-5}\,\mathrm{mm}$
Hg (in case 1), then $n_2/n_1 = 5 - 6$. At a higher pressure
$(n_2/n_1 > 5 - 6)$, no current disruption occurs [6.10].

6.6   Plasma Stabilization (Raising of Thresholds) of Electron–Ion Beam Instabilities and of the Pierce Instability

The beam instability in the experiments discussed in sections 6.3–6.5 was
caused by the interaction of two components of the system: beam particles
and one species of particles of a fixed medium (either ions or electrons). In
this section we will consider the effect of the third component, that is, of
the electrons of the 'excess' plasma injected into the quasineutral electron
beam, on the thresholds of the electron–ion instabilities and of the Pierce
instability.

Experiments on studying the thresholds of a three-component plasma
were carried out with the experimental set-up schematically shown in
figure 6.1. The plasma density was varied by controlling the pressure of
the neutral gas (ionized by the electron beam itself). The more essential
results are shown in figures 6.33 and 6.34.

Figure 6.33 shows the limiting electron beam current as a function
of plasma electron density (the current was restricted by the beam–drift
instability and the density was measured by the differential anode method).
The figure shows two curves, corresponding to two different beam radii:
0.5 cm (curve 1) and 0.25 cm (curve 2). We see that in the case of curve 1,
the threshold of the beam–drift instability increases considerably beginning
at a certain nitrogen pressure (i.e., at a certain plasma density; see also

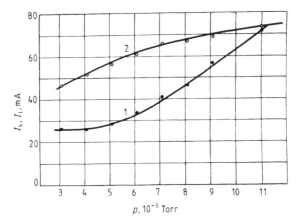

Figure 6.34   Threshold current I_s of excitation of axially symmetric electron–ion oscillations, 1, and the limiting beam current I_{lim}, 2, as functions of nitrogen pressure; $W_1 = 200\,\text{eV}$, $H = 4000\,\text{Oe}$, $L = 60\,\text{cm}$ [6.10].

figure 6.34); this effect is not observed in beams of sufficiently small radius.

Figure 6.34 shows how the threshold I_s of the Budker–Buneman instability increases with increasing plasma density (controlled by varying the nitrogen pressure). We see that at a certain plasma density corresponding to the nitrogen pressure $p \simeq 1 \times 10^{-5}\,\text{mm Hg}$, the threshold I_s becomes equal to the limiting beam current (curve 2 in figure 6.34) and the Budker–Buneman instability is not observed any more. Figures 6.33 and 6.34 thus illustrate the effect of plasma stabilization of electron–ion beam instabilities. Thus the thresholds of the two electron–ion instabilities are unequally sensitive to plasma density variations (to gas pressure changes): if $p \geqslant (1 - 1.2) \times 10^{-5}\,\text{mm Hg}$, only the beam–drift instability can actually be observed.

Switching now to the theoretical interpretation of the data obtained, let us look again at the experimental conditions [6.18]: beam length $L = 10 - 30\,\text{cm}$, $k_z = n\pi/L$; $n = 1, 2, 3, \ldots$, thermal velocity of plasma electrons $v_e = (0.5 - 2) \times 10^8\,\text{cm}\,\text{s}^{-1}$, frequency of oscillations generated in the instability $\omega = 6\pi \times 10^5\,\text{s}^{-1}$. These are conditions under which the so-called hot plasma approximation is realized:

$$\omega \ll k_z v_e. \tag{6.25}$$

In this approximation, it is necessary to write the dispersion equation (3.30) of oscillations, taking into account the thermal motion of electrons in the plasma. This treatment yields the following result (see, e.g., [6.12, 6.18]): the second term in (3.30) takes the form

$$\omega_p^2(k_z^2/k^2)/(\omega^2 - k_z^2 v_e^2).$$

In the approximation of (6.25), this term reduces to a constant $(k^2 r_D^2)^{-1}$, where $r_D = v_e/\omega_p$ is the electron Debye radius.

Transferring this constant to the right-hand side of equation (3.30), we obtain the expression $[1 + (k^2 r_D^2)^{-1}]$ by which the instability threshold will obviously be multiplied. We therefore obtain that the threshold for the beam–drift instability (see section 3.4) is

$$I_c \simeq \frac{ma^2}{4e} k^2 u^3 \left[1 + (k^2 r_D^2)^{-1}\right] \left[1 + \frac{2Su}{a^2 \omega_{H_e} k_z}\right]^{-1} \quad (6.26)$$

and for the Budker–Buneman instability (see section 3.2), it is

$$I_s = \frac{ma^2}{4e} k^2 u^3 \left[1 + (k^2 r_D^2)^{-1}\right] \left[1 + \left(\frac{m}{M} \frac{k^2}{k_z^2}\right)^{1/3}\right]^{-3}. \quad (6.27)$$

Since the Pierce instability threshold is obtained from the Budker–Buneman threshold in the limit $M \to \infty$, we find for the Pierce threshold from (6.27) that

$$I_P = \frac{ma^2}{4e} k^2 u^3 \left[1 + (k^2 r_D^2)^{-1}\right] = \frac{ma^2}{4e} k^2 u^3 \left(1 + \frac{4\pi e^2}{k^2 T_e} n_2\right). \quad (6.28)$$

We see that expressions (6.26)–(6.28) contain effects of stabilization of electron–ion instabilities and of the Pierce instability. These effects are due to an increase in the contribution of plasma electrons to the dielectric permittivity of the medium at low oscillation frequencies which result in an additional factor

$$\left[1 + (k^2 r_D^2)^{-1}\right] = \left(1 + \frac{4\pi e^2}{k^2 T_e} n_2\right)$$

in (6.26)–(6.28). Under the conditions of the experiments we are discussing now, the value of $(k^2 r_D^2)^{-1}$ at a sufficiently high gas pressure $(p = 7 - 8) \times 10^{-5}$ mm Hg) and beam radius $a \simeq 1/k \simeq 0.5$ cm is about several tens. Therefore, as the plasma density increases, the experimentally observed sharp increase in the limiting current (see figure 6.34) is in qualitative agreement with the theoretical relation (6.26) for the beam–drift instability threshold and with relation (6.28) for the Pierce instability threshold. For a beam of half the radius $(a = 0.25 \text{ cm})$, the term $(k^2 r_D^2)^{-1}$ in the numerator of relation (6.26) is four times smaller. A weak dependence of the limiting current on plasma density also appears as qualitatively natural in this case. When the thresholds (6.26)–(6.28) are evaluated quantitatively, it is necessary to take into account the fact that the mean velocity of beam electrons under these conditions (see figure 6.33) is considerably smaller (almost by a factor of two) than the initial velocity not perturbed by the beam–plasma interaction [6.18]. In view of this, there is a quantitative agreement between the theoretical values of the thresholds (6.26)–(6.28) and the experimental data.

Unequal sensitivity of excitation thresholds of the two instabilities, discussed above, to adding an excess of plasma to a quasineutral beam (see figures 6.33 and 6.34) is given the following explanation. I have mentioned already (see section 4.5) that the 'hydrodynamic' Budker-Buneman instability is stabilized by the velocity spread of beam electrons. In contrast to this, the beam–drift instability (sustained also by the convective mechanism, in addition to the purely beam mechanism) is not stabilized by the beam velocity spread (see section 3.4); the beam–drift instability exists for a wider range of variation of plasma density (gas pressure)—see figures 6.33 and 6.34.

We thus find that the effects of plasma stabilization of the beam–drift electron–ion and Pierce instabilities are clearly observed in experiments and offer an additional corroboration of the existing linear theory. The following remarks are needed to conclude this section. The beam current in the experiments described here was restricted by the capabilities of the 'electron gun', determined by the Child–Langmuir law for the accelerating gap between the gun's cathode and the accelerating electrode (the grid). This restriction was removed in another series of experiments where beam-plasma discharge and a hot cathode were used. The effect of 'excess' plasma on the limiting beam current is described for these conditions in section 6.9 which deals with the properties of plasma beams.

6.7   Limiting Currents of Ion and Relativistic Electron Beams

At the present moment, almost no data are available on limiting currents in ion beams. Paper [6.19] mentions that current disruption (formation of a virtual cathode) may occur in beams of ions of heavy elements (lead, thallium) with 1 keV energy when the current density exceeds 100 mA/cm². The authors of [6.19] are of the opinion that this phenomenon is in qualitative agreement with the theory of aperiodic instability of quasineutral ion beams, as outlined in section 2.3. It remains unclear to what extent the beam in the experiments [6.19] was quasineutral (or over-compensated) and what the excess plasma density was.

The transmission of a high-intensity ion beam (with 70 keV ion energy) through high-density plasma is described in [6.21]. The beam current (5 kA) exceeded by three orders of magnitude the Bursian maximum beam current without compensation by electrons. Free passage of the beam was observed when plasma density ($n_e = 5 \times 10^{13}$ cm^{-3}) was greater by two orders of magnitude than the beam density. No experimental data on quasineutral ion beam are reported in [6.21].

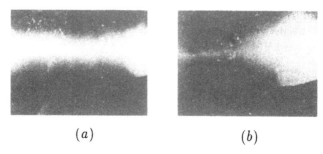

(a) (b)

Figure 6.35   Propagation of a relativistic electron beam
through a gas (air at a pressure 0.2 mm Hg). The beam travels
from right to left, the length of the segment shown in the figure
is 40 cm: (a) beam electron energy $W_1 = 2.5\,\text{MeV}$, $I = 20\,\text{kA}$,
the beam propagates easily; (b) $W_1 = 1.5\,\text{MeV}$, $I = 40\,\text{kA}$, the
beam shuts off [6.20].

Experiments with quasineutral ion beams thus remain a matter of future
effort.

Experimental data on limiting currents of (pulsed) relativistic beams in
the absence of charge neutralization and in a strong longitudinal magnetic
field are given in section 6.1. There are no definitive experimental data
yet on quasineutral relativistic beams†. As for the fact of shutting of a
relativistic beam by its own space charge, it is well illustrated by figure 6.35
which represents the photograph of the beam shown in [6.20].

Beam shutting under the conditions of figure 6.35 occurs at a current
which is several times greater than the maximum current (2.27) and (2.30)
in a beam without ions. However, the degree of neutralization of the beam
space charge in the experiments [6.20] remained undetermined.

In conclusion, it is necessary to point out that, as I have already
mentioned in the Introduction, relativistic beams are used nowadays to
create very high-power generators of HF electromagnetic waves. These
generators employ a plasma which serves two functions: first, it provides
neutralization of the beam space charge (without it, the limiting beam
current may be lower than the generation threshold); second, the plasma
makes it possible to generate electromagnetic waves in new frequency
ranges [6.22, 6.23]. For such applications, it is of primary importance
to know the thresholds of beam instabilities and the limiting beam
currents (e.g., to know how they grow with increasing plasma density; see
section 6.6).

---

† So far experiments with relativistic beams have been conducted in only two
asymptotic situations: either with no compensation of the space charge of the
beam, or with an 'excess' plasma in the beam, at a density higher than the beam
density. In the latter case, the limiting beam currents are so large (see section 6.6)
that apparently no current disruption occurs.

6.8 On the Efficiency of Application of Quasiclassical Approximation to an Analysis of Large-Scale Beam Instabilities

A detailed comparison of the experimental data with the theory of beam instabilities, given in this chapter, demonstrates that the quasiclassical approximation used in this book 'works' quite well in the case of large-scale instabilities (at $k_z \simeq \pi/L$, $k_\perp \simeq 1/a$), that is, far beyond the limits of its strict applicability: indeed, it would appear to be valid only for small-scale perturbations, when $k_z \gg \pi/L$, $k_\perp \gg 1/a$. I will present now the arguments in favour of this approach since the applicability of this approximation to large-scale instabilities is sometimes criticized. The first aspect to be emphasized is that owing to the simplicity and elegance of the quasiclassical approximation, it allows one to derive immediately explicit relations between instability thresholds and all the parameters of the system. As was shown in detail in this chapter, these relations yield a clear and detailed experimental program for thoroughly testing the theory, and this program has indeed been implemented in the experiments described above. On the other hand, the use of a more rigorous but considerably less transparent theory (which, as became clear later, gives virtually the same results) would be unlikely to generate in experimenters enough enthusiasm for starting such a program. I have no doubt that readers know a number of examples from other fields of physics in which a theory, which appeared to be valid only qualitatively or within an order of magnitude, actually provided accuracy much better than had been expected. Therefore, having gained the knowledge of the actual progress 'from the horse's mouth', we can state firmly that the quasiclassical approach proved quite fruitful and played a radically heuristic role in the physics of particle beams.

It should be added to the above that in contrast to instability threshold experiments, experiments on measuring the limiting currents in beams should be judged in terms of a non-linear theory, and in this sense it is immaterial what approximation of the linear theory is employed: any one of them are invalid in the strict sense of the word. The fact that the interruption of beam current is (unexpectedly) well described (and predicted) by the quasiclassical approximation of the linear theory is another argument in favour of its creative usefulness.

6.9 Instability of Plasma Beams. Formation of the Virtual Cathode in Plasma Beams

We will discuss here instabilities of a three-component system in which the beam electron concentration is usually small compared with the concentration of secondary (plasma) particles: electrons and ions. These instabilities are usefully classified under two groups by the type of their effects on the beam itself.

The first group comprises those instabilities which result in a drastic restructuring of the beam (we refer to this phenomenon as the destruction of the plasma beam). This process starts with a seemingly quite 'innocuous' centrifugal instability, which changes the parameters of the plasma beam in such a way that the drift generation of ion sound becomes possible. In its turn, the ion sound affects the beam in such a manner that it creates the initial conditions for triggering the beam–drift instability. This last instability produces a dynamic shutting of the beam, that is, the formation of an oscillating virtual cathode (for details, see [6.24, 6.25]).

However, if the initial parameters are chosen so that the centrifugal and ion-sound instabilities do not arise, the experiment demonstrates (in good agreement with the theory) that the beam is stable with respect to the beam-drift instability and there is no beam shutting. In the process, second-group (microscopic) instabilities develop 'quietly' in the beam, not threatening the existence of the plasma beam itself and manifesting themselves in the generation of the high-frequency electron oscillations and waves (including Langmuir waves) discussed above. For this reason, we consider here the first group of instabilities.

The experimental analysis of these instabilities [5.15, 5.28, 5.29, 6.24, 6.25] was initiated in order to answer the following question: to what extent does the limiting current of a (quasineutral) electron beam change if additional plasma of sufficiently high density is introduced into the beam? In other words, is it possible to practically remove the limitation on the beam current and produce a beam whose current is, say, a hundred times higher than the Pierce threshold (section 2.2.1)?

The experiments used a plasma source (figures 5.3, 5.4) in which the electron beam current could be greater by two orders of magnitude than in experiments with electron gun (of the type of figure 2.1). These experiments showed that at least two distinct regimes are observed in the propagation of the electron beam.

First regime. If the relative plasma concentration, $\alpha = n_2/n_1$, exceeds a certain critical value α_c, the beam in the plasma can be said to be macroscopically stable: all electrons that have left the cathode freely traverse the plasma and reach the beam collector.

Second regime. If $\alpha < \alpha_c$, the beam is strongly unstable: electric fields of very high amplitude arise, whose azimuthal component reaches $E_\varphi \simeq W_1/ea$, where W_1 is the energy of beam electrons. A large number of electrons leaving the cathode then fail to reach the beam collector; they drift rapidly in these electric fields (crosswise to the longitudinal magnetic field) and are deposited on the walls of the discharge chamber. This process can be characterized by a diffusion

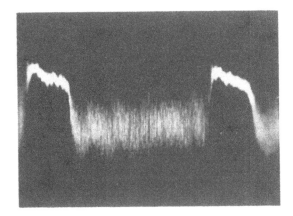

Figure 6.36 Oscilloscope trace of the anode electron current (upward deflection corresponds to increasing electron current): long period $T = 500\,\mu s$, short period $\tau = 0.15\,\mu s$, the double amplitude of HF oscillations 2A. Discharge current $I_d = 1.8\,A$, $W_1 = 180\,eV$ [6.24].

coefficient which is much greater than Bohm's anomalous diffusion coefficient (see [2.8]).

In the latter state of the beam, a virtual cathode is formed (see figure 6.36) which pulses with time, and the average beam current drops sharply [2.8, 5.25], see figure 6.37. In addition to these two states of the beam, at $\alpha \lesssim \alpha_c$ there exists the third, intermediate state [6.25, 5.29].

Experiments [6.24, 6.25] demonstrated that the critical quantity α_c is determined by the beam electron energy and by the temperature of plasma electrons

$$\frac{n_{2c}}{n_1} \equiv \alpha_c \simeq 4\frac{u}{v_2} \tag{6.29}$$

that is, the condition of the macroscopic beam instability has the form

$$j > j_e = \frac{en_2v_2}{4} \tag{6.30}$$

or

$$n_1 u > \frac{n_2 v_2}{4}. \tag{6.31}$$

Therefore, the limiting beam current density is equal to the density of the random current of plasma electrons.

Let us look at a numerical example. Assume $n_2 = 1 \times 10^{12}\,cm^{-3}$, $W_1 = 120\,eV$, $T_e = 3\,eV$ and a beam 1 cm in diameter; according to (6.29), the limiting beam current is then 4 A. Let us compare this current with the Pierce threshold. If the beam passes through a tube 10 to 30 cm in diameter, the Pierce current (2.18) is $I_p \simeq 25\,mA$.

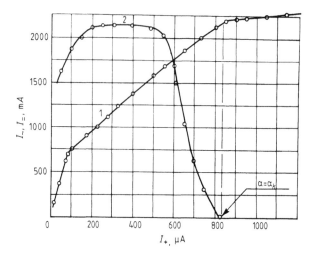

Figure 6.37   DC component, 1, and oscillation amplitude, 2, of the beam current as functions of plasma concentration n_+. The ion current I_+ to the collector of the differential anode is the indicator of n_+. $W_1 = 120\,\text{eV}$, $H = 1800\,\text{Oe}$, $L = 85\,\text{cm}$, $p = 5 \times 10^{-6}$ mm Hg [6.24].

We conclude that the limiting current of primary electrons in a stable high-density plasma beam exceeds the Pierce current by more than two orders of magnitude,

$$\frac{I_{\lim}}{I_P} \mathrel{\gtrsim} \frac{n_2}{n_1} \tag{6.32}$$

and grows in approximate proportion to the plasma density.

The factor that causes the current disruption and virtual cathode formation in a plasma beam is the beam-drift instability. It also manifests itself in the generation of electric fields in which plasma ions are accelerated to energies which lie close to, or are even greater than, W_1 (figure 6.38); this is why the plasma beam can be used as an efficient injector of hot plasma into traps with magnetic mirrors [5.34] (see section 7.1).

Condition (6.31) of beam-drift instability turns out to be connected with the formation of electric double layers in the plasma (see section 9.2).

It is important to point out that the condition of beam-drift instability (6.30) and (6.31), which is well satisfied in the beam propagation region, does not contradict condition (5.21) of stable discharge in the source of the plasma beam where the (local) plasma density is much higher ($\alpha > \alpha_c$). For this reason, if there is a longitudinal plasma density gradient, a stable hot-cathode discharge (with $\alpha > \alpha_c$ in the immediate vicinity of the cathode) can produce a plasma beam which is unstable with respect to the formation of a virtual cathode far from the source (where $\alpha < \alpha_c$).

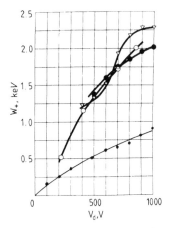

Figure 6.38   Maximum energy W_m of Larmor rotation in
various gases as a function of beam electron energy ($W_1 = eV_d$).
The average ion energy is about $\frac{1}{2}W_m$. The beam current is
(1–1.5) A, in the regime of oscillating virtual cathode. $\bigcirc$, H_2;
∇, N_2; $\bullet$, He; $\cdot$, Xe. In the case of H_2, He and N_2, $H = 4000$ Oe,
and in the case of Xe, $H = 1000$ Oe [5.34].

6.10   Experimental Observation of Diocotron Instability

A clear description of the manifestations of the *diocotron* instability in a
hollow (non-compensated) relativistic electron beam were given in [6.27].
The electron energy was 1.9 MeV, the beam current reached up to 90 kA,
and the pulse duration was 60 ns. The beam diameter is 1.8 cm but its
thickness is only 300 μm; it propagates close to the walls of a metal tube
along a strong external magnetic field of 20 to 90 kG†. Azimuthal drift
electron fluxes arise in a radial, strongly non-uniform electric field of the
space charge of the beam at right angles to a strong magnetic field; in
agreement with Rayleigh's criterion, the transverse velocity profile of these
fluxes is found to be unstable (see section 3.6). The instability triggers the
generation of vortex structures whose photograph is shown in figure 6.39.
A comparison with the results of other authors shows [6.27, 6.28] that such
structures exist in a very wide range of experimental conditions: the beam
current varies by at least 9 orders of magnitude. This is not surprising since
the diocotron instability has no beam current threshold.

Splitting of a relativistic beam into filaments as a result of the diocotron
instability was observed experimentally in [6.29] under conditions of the
beam current being close to the limiting (Bursian) value.

---

† Such a high beam current should not cause any surprise: the potential depression
in the beam is proportional to squared beam thickness, and this is a very low
quantity. Besides, the limiting beam current may be very high since the beam
propagates quite close to a metal wall.

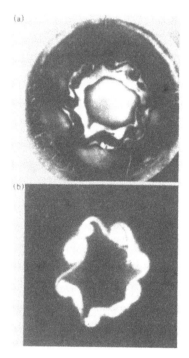

Figure 6.39   Electron beam vortices, seen as imprints on a
graphite collector. (*a*) Electron beam of energy 1.9 MeV and
current 90 kA, having annular cross section with very small
ring thickness. (*b*) Low-energy electron beam, beam current
58 μA. The results are reproduced from [6.27]. We see that
vortex structures are observed in the range of current variation
of a hollow tubular beam by 9 orders of magnitude.

An interesting analysis of the diocotron instability, revealing high-
intensity electron vortices, was reported in a series of papers [6.30–6.32]
(see also the literature cited therein). An experimental investigation of
stabilization of this instability was described in [6.32].

On a new classification of instabilities of non-uniform non-compensated
electron beams and on the possibility of observation of picturesque
manifestations of these instabilities, see [3.21–3.27].

References

[6.1]   Atkinson H H 1963 *Abstr. 5th Ann. Meet. Amer. Phys. Soc.* Div.
Plasma Phys., San Diego, Calif. (USA)

[6.2]   Nezlin M V and Solntsev A M 1967 *Zh. Eksp. Theor. Fiz.* **53** 437
(Engl. Transl. 1968 *Sov. Phys.–JETP* **26** 290)

[6.3] Nezlin M V, Taktakishvili M I and Trubnikov A S 1971 *Zh. Eksp. Theor. Fiz.* **60** 1012 (Engl. Transl. 1971 *Sov. Phys.–JETP* **33** 548)

[6.4] Mkheidze G P, Pulin V I, Raizer M D and Tsopp L E 1972 *Zh. Eksp. Theor. Fiz.* **63** 103 (Engl. Transl. 1973 *Sov. Phys.–JETP* **36** 54)

[6.5] Abrashitov Yu I, Koidan V S, Konyukov V V *et al* 1974 *Zh. Eksp. Theor. Fiz.* **66** 1324 (Engl. Transl. 1974 *Sov. Phys.–JETP* **39** 647)

[6.6] Read M E, Nation J A 1975 *J. Plasma Physics* **13** 127

[6.7] Miller R B, Straw D C 1977 *J. Appl. Phys.* **48** 1061

[6.8] Voronin V S, Krastelev V G, Lebedev A N, Yablokov B N 1978 *Fizika Plasmy* **4** 604 (Engl. Transl. *Sov. J. Plasma Phys.* **4** 336)

[6.9] Kravchuk V N, Kondratenko A N 1987 *Zh. Tekhn. Fiz.* **57** 74 (Engl. Transl. *Sov. Phys. Tech. Phys.* **32** 44)

[6.10] Nezlin M V , Taktakishvili M I and Trubnikov A S 1968 *Zh. Eksp. Theor. Fiz.* **55** 397 (Engl. Transl. 1969 *Sov. Phys.–JETP* **28** 208)

[6.11] Volosov V I 1962 *Zh. Tekhn. Fiz.* **32** 566 (Engl. Transl. *Sov. Phys. Tech. Phys.* **7** 412)

[6.12] Nezlin M V 1981 *Fizika Plasmy* **7** 1048 (Engl. Transl. 1981 *Sov. J. Plasma Phys.* **7** 575)

[6.13] Vladyko V B, Rudyak Yu V and Rukhlin V G 1987 *Fizika Plasmy* **13** 1246 (Engl. Transl. 1987 *Sov. J. Plasma Phys.* **13** 720)

[6.14] Hirose A, Paulson J D, Scarsgard H M and Wolfe S 1983 *Phys. Rev. Lett.* **51** 1179

[6.15] Nezlin M V, Sapozhnikov G I and Solntsev A M 1966 *Zh. Eksp. Theor. Fiz.* **50** 349 (Engl. Transl. 1966 *Sov. Phys.–JETP* **23** 232)

[6.16] Bogdankevich L S, Raizer M D, Rukhadze A A and Strelkov P S 1970 *Zh. Eksp. Theor. Fiz.* **58** 1219 (Engl. Transl. 1970 *Sov. Phys.–JETP* **31** 655)

[6.17] Gabovich M D and Pasechnik L L 1959 *Zh. Eksp. Theor. Fiz.* **36** 1025 (Engl. Transl. 1969 *Sov. Phys.–JETP* **9** 272)

[6.18] Nezlin M V, Taktakishvili M I and Trubnikov A S 1976 *Zh. Tekhn. Fiz.* **46** 64 (Engl. Transl. 1976 *Sov. Phys. Tech. Phys.* **21** 34)

[6.19] Grishin S D, Yerofeev V S and Zharinov 1973 *Plasmennye Uskoriteli (Plasma Accelerators)* (Moscow: Mashinostroyeniye) p. 54

[6.20] Uglum J R, McNeill W H, Graybill S E and Nablo S V 1969 *Proc. 9th Intern. Conf. on Physics of Ionized Gases* Bucharest p. 574

[6.21] Nakagawa Y 1985 *Phys. Fluids* **28** 1956

[6.22] Kuzelev M V, Rukhadze A A, Strelkov P S and Shkvarunets A G 1987 *Fizika Plasmy* **13** 1370 (Engl. Transl. 1987 *Sov. J. Plasma Phys.* **13** 793)

[6.23] Fainberg Ya B 1985 *Fizika Plasmy* **11** 1398 (Engl. Transl. 1985 *Sov. J. Plasma Phys.* **11** 803)

[6.24]   Nezlin M V 1961 *Zh. Eksp. Theor. Fiz.* **41** 1015 (Engl. Transl. 1962
         Sov. Phys.-JETP **14** 723)

[6.25]   Nezlin M V 1967 *Zh. Eksp. Theor. Fiz.* **53** 1180 (Engl. Transl. 1968
         Sov. Phys.-JETP **26** 693)

[6.26]   Nezlin M V, Taktakishvili M I and Trubnikov A S 1970 *Zh. Tekhn.
         Fiz.* **40** 392 (Engl. Transl. 1970 *Sov. Phys. Tech. Phys.* **15** 292)

[6.27]   Peratt A L and Snell C M 1985 *Phys. Rev. Lett.* **54** 1167

[6.28]   Chen K W, Kim S H and McKinley M C 1987 *Phys. Fluids* **30** 3306

[6.29]   Fedotov A V and Shkvarunets A G 1987 *Fizika Plasmy* **13** 1068
         (Engl. Transl. 1987 *Sov. J. Plasma Phys.* **13** 614)

[6.30]   Kervalishvili N A 1989 *Fizika Plasmy* **15** 753 (Engl. Transl. 1989
         Sov. J. Plasma Phys. **15** 1136)

[6.31]   Driscoll C F and Fine K S 1990 *Phys. Fluids* B2(6) 1359

[6.32]   Rosenthal G, Dimonte G and Wong A Y 1987 *Phys. Fluids* **30** 3257

7 Acceleration and Heating of Plasma Ions in a Regime Close to the Limiting Beam Current

7.1 Heating of Ions in Unstable Plasma Beams. An Unstable Plasma Beam as an Injector of Hot Ions into a Plasma Trap with Magnetic Mirrors

A profound relation between beam current disruption and acceleration (heating) of plasma ions was discovered in the experiments reported in [5.15, 5.25, 5.34, 6.10]. The magnetic field in these experiments was either uniform or was shaped into a trap with magnetic mirrors; either protons or lithium ions were used as plasma ions. The experimental setup used in the experiments with protons is shown in figure 7.1. A stainless steel chamber, 30 cm in diameter and 2 m long, was evacuated to a pressure of $(2-3) \times 10^{-7}$ mm Hg.

The first experiments were carried out in a uniform field of strength up to 7800 Oe. A large part of the experiments were conducted in a trap

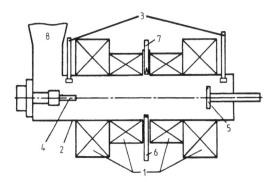

Figure 7.1 Schematic diagram of the experimental setup [5.34]: 1, coils creating the magnetic field: 2, vacuum chamber; 3, titanium sorption vacuum pumps; 4, plasma source; 5, anode (plasma beam receptacle); 6, analyser probe; 7, receiver for fast neutral atoms; 8, pump connection pipe.

with magnetic mirrors, with the field at the center (H_0) up to 7200 Oe and the field in the mirrors (H_{max}) up to 9600 Oe at the mirror ratio $H_{max}/H_0 = 1.33$ and the mirror-to-mirror spacing 100 cm.

The plasma beam was produced using the source described in chapter 5 (see figure 5.3). The beam propagated along the axis of the vacuum chamber and was received at the anode which was placed at a distance of 160 cm from the source. The cathode diameter in the source was 1 cm, the diameter of the hole in the discharge chamber was 2 cm. If 200 cm^3 of hydrogen was fed into the source per hour, the hydrogen pressure in the discharge chamber was 10^{-3} mm Hg and the plasma density in the beam was of the order of 10^{12}–10^{13} cm^{-3}. In experiments with the magnetic trap field, the plasma source was placed outside the trap; the distance from the output orifice of the discharge chamber to the magnetic mirror (i.e., to the point of maximum field) was 25 cm.

The energy of ions outside the plasma beam and their concentration were measured using methods described in section 5.6.

Two discharge regimes were used in the experiments: the continuous regime at a voltage V_d up to 1 kV and current I_d up to 4 A, and the pulsed regime at a voltage up to 2 kV, current up to 40 A and the discharge pulse duration $T_d = 0.5$-10 ms. The plasma beam, which was the object of our study just as in [5.25], consisted of three components: (1) fast primary electrons emitted by the heated cathode and accelerated in the layer of the cathode potential drop to energies up to 0.5–1 keV, (2) plasma electrons with energies of several electron volts and (3) plasma ions. The initial ratio of plasma density to the density of primary beam electrons was several tens. As was shown in chapter 6, such a plasma beam is unstable under certain conditions with respect to the formation of a virtual cathode in the stream of primary electrons. In order to study the possible relationship between this kind of instability of the plasma beam and the heating of ions, we measured the energies of Li$^+$ ions ejected from the plasma beam transversely to the magnetic field. First experiments were carried out in a setup with uniform magnetic field and lithium plasma beam. The ion energies were measured by means of the retarding potential method, using a three-electrode cylindrical analyser which enveloped the plasma beam. The measurements showed that fast Li$^+$ ions (with energies of the order of several hundred eV) appear in the plasma beam only in the regime in which the beam is unstable with respect to the formation of a virtual cathode. In the stable beam regime, the energies of Li$^+$ ions do not exceed 20–30 eV. Figure 7.2 shows, in the upper right-hand corner, the oscillograms of the electron current to the anode (upper curve) and the ion current I_{col} to the analyser's collector at $V_{col} = +250$ V (lower curve). The parts of the upper oscillogram with high-amplitude oscillations correspond to an unstable regime, similar to the regime of figures 6.36 and 6.37, in which, as was shown in section 6.9, a virtual cathode oscillates in the beam; the relatively smooth segments correspond to the 'stable' regime. We see

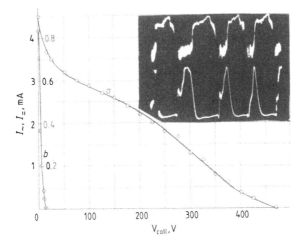

Figure 7.2 Retarding-field current–voltage characteristics of the analyser of accelerated ions in the (*a*) unstable and (*b*) stable states of the plasma beam. The inset shows oscillograms of the electron current at the anode (upper curve, downward deflection corresponding to decreasing current) and the current of accelerated ions at the analyser's collector (lower curve, upward deflection corresponding to increasing current). Scanning time, 1 ms [5.34].

that the fast Li^+ ions emerge from the plasma beam only in the unstable regime. Figure 7.2 shows, in addition to oscillograms, the retarding curves which characterize the energy spectrum of plasma ions in the unstable and the stable states of the plasma beam. It was shown in section 6.9 that strong rapidly varying electric fields of amplitude up to 200 V/cm arise in plasma beams which are unstable with respect to the formation of a virtual cathode. The frequency spectrum of these fields is found to be continuous and includes the Larmor frequency of plasma ions. Multiple stochastic heating of plasma ions can occur in fields with such spectra; it corresponds to the fact that heated ions have a continuous energy spectrum. In view of the experimental data given above, we have to assume that the observed efficient heating of ions does take place in these fields.

Experiments with a hydrogen plasma beam, conducted in a uniform magnetic field, gave results which are quite similar to those obtained with the lithium beam. They have also demonstrated that almost all ions entering the plasma beam from the source are involved in the heating, and that the predominant part of the energy of heated ions (W_+) is connected with rotation perpendicularly to the magnetic field. This type of heating of ions is favourable for their accumulation in a trap with magnetic mirrors. The main purpose of all subsequent experiments was in fact an investigation of this possibility.

In choosing the particular regime of heating ions in the plasma beam

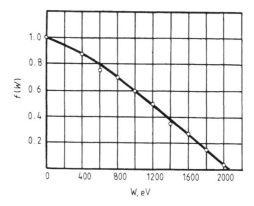

Figure 7.3 Integral energy spectrum of ions in magnetic trap: $W_1 = 1.1\,\text{keV}$, $I_d = 12\,\text{A}$, $I_a = 2.6\,\text{A}$, $n_{+,\text{max}} = 7.5 \times 10^{10}\,\text{cm}^{-3}$ [5.34].

special attention was paid to the case in which the flux of heated ions ejected from the beam was most intense. For this reason, the experiment used a plasma source with increased volume of the discharge chamber; the source was operated in high-power pulsed discharge mode. The discharge pulse duration was varied in the range from 0.5 to 10 ms. The data given below were obtained for $T_d = 0.5\,\text{ms}$.

Figure 7.3 shows a typical retarding-field current–voltage characteristic of the analyser probe placed close to a wall of the vacuum chamber. Shown on the abscissa axis is the positive collector potential, and the ordinate is a quantity proportional to the total charge q passing through the collector probe during the discharge pulse. Curve 7.3 directly characterizes the integral energy spectrum of ions. We see that one half of the ions recorded by the probe have energies in the range $(1\text{--}2)\,\text{keV} \leqslant W_+ \leqslant (2\text{--}2.2)\,\text{keV}$. We also see that the average ion energy W_{av} is approximately equal to one half of the maximum ion energy W_{max} and, therefore, the quantity W_{max} (which is readily measurable) characterizes the energy distribution of ions quite unambiguously. In estimates of fast ion density given below, it is assumed that

$$W_{\text{av}} \simeq \tfrac{1}{2} W_{\text{max}} \simeq 1\,\text{keV}. \tag{7.1}$$

Quite similar results were obtained in experiments in which ions of heavier gases were heated: helium, argon, nitrogen, krypton and xenon. Figure 6.38 shows the energy W_{max} of the Larmor rotation of hydrogen, helium, nitrogen and xenon ions as a function of accelerating (discharge) voltage of the plasma beam source.

It is thus clear that in a plasma beam which is unstable with respect to the formation of a virtual cathode in the flow of primary electrons, ions are very efficiently heated up, so that the beam can be used as a high-power injector of fast ions for magnetic traps. A beam 2 m long with primary

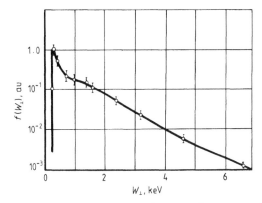

Figure 7.4 Energy spectrum of protons in a magnetic trap. W_1 is the energy of motion of ions perpendicularly to the magnetic field. $W_{av} = 1.2$ keV, beam current $I_d = 10$ A, $H = 3000$ Oe, $W_1 = eV_d = 1$ keV [2.8].

electron energy of about 1 keV and current 20 A yields a quasistationary proton current of about 2 A at energies 1 to 2 keV. For the proton lifetime of $\tau \simeq 50$ ms, this leads to a plasma with fast proton density of about 10^{11} cm^{-3} in a trap of 10^4 cm^3 volume.

The phenomenon of ion heating in unstable plasmas is used in well-known experiments in which the plasma is studied in magnetic traps with combined magnetic fields [7.1, 7.2]. Figure 7.4 shows the energy spectrum of protons which fill one such trap as a result of an instability of the plasma beam passing along its axis (the spectrum is reproduced from [7.3]). We see that the spectrum contains a considerable number of ions with energies of several keV, that is, energies several times greater than that of primary electrons of the beam ($W_1 = 1$ keV); the average ion energy (their temperature) is 1.2 keV. (Note that particles with energies as high as 25–30 keV are found among the accelerated (heated) ions, although their number is relatively low.)

It is thus found to be not only feasible but also convenient to produce pure high-temperature plasma using an unstable plasma beam sent along the axis of a trap with magnetic mirrors. An experimental investigation of the properties of such plasmas in traps yielded results which are definitely of interest in connection with the problem of controlled fusion [7.2].

As we see from the contents of this subsection and of section 6.9, the main reason for the efficient heating of ions in a high-density plasma beam is the beam–drift instability. This instability develops if the initial plasma density exceeds that of primary electrons by not more than two orders of magnitude. For example, the criterion (6.30), (6.31) implies that for $W_1 = 1$ to 1.5 keV and $T_e = 2$ eV, the ratio $\alpha = n_2/n_1$ must not exceed 100.

Note that heating of plasma ions owing to the instability of the

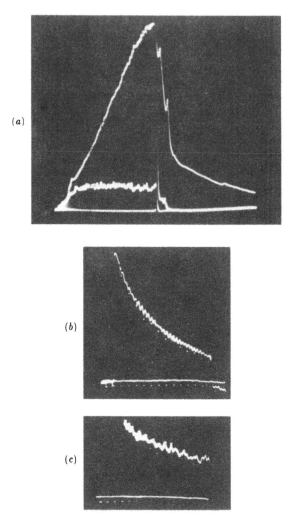

(a)

(b)

(c)

Figure 7.5   Oscillograms of the flux of fast charge-exchange
neutrals (regarded as a plasma density indicator) from the time
the plasma beam was turned on and with a time delay by τ
with respect to this time ((a), scan time 3 ms, $\tau = 0$; (b), scan
time 30 ms, $\tau = 3.6$ ms; (c), scan time 30 ms, $\tau = 7$ ms). The
lower curve in figure (a) was obtained with the stabilizing field
(creating the 'B minimum') turned off. The disruption of the
signal at the end of figure (b) is connected with turning off of
the stabilizing field (for details, see [1.17]).

and a field shaped into a trap with mirrors) was also observed in the
experiments (and used successfully to create hot fusion plasma) in [7.4,
7.5]. Similar phenomena were observed in [7.6, 7.7] and also in relativistic
beams [7.8].

beams [7.8].

Another interesting phenomenon was discovered in the experiments [5.34, 7.3] on the accumulation of hot ions in a trap with magnetic mirrors (this time, with the 'B minimum' at the center of the trap [7.1, 7.2]). It was found that under certain conditions, the process of accumulation of the plasma injected into the trap from an unstable plasma beam propagating along its axis and the subsequent decay of the plasma (after the beam has been turned off) looks as shown in figure 7.5(a). We see that the decay is not smooth but is accompanied with abrupt drops. An analysis of this phenomenon has demonstrated [7.2] that these drops indicate that the plasma escapes owing to the so-called *negative-mass instability*. This instability arises because the period of rotation of particles in magnetic field increases as their (transverse) energy increases. As a result of instability, waves are generated at a frequency slightly higher than the cyclotron frequency. The reason for the decrease in the frequency of rotation of particles in magnetic field in response to increasing particle energy is, in this case, the fact that the lower the (transverse) energy of particles, the farther they penetrate into the mirrors (i.e., into the region of higher magnetic field) and the higher their Larmor frequency. There exists a very different example of a physically similar situation: the frequency of rotation of electrons in a magnetic field decreases owing to the relativistic increase in their mass. The larger the acceleration of a particle (e.g., in a cyclic accelerator), the slower its rotation is (hence the name 'negative-mass effect'). As we know [7.9], this situation leads to phase and spatial bunching of particles and to the generation of wave perturbations at a frequency slightly higher than the cyclotron frequency. For instance, such highly efficient HF emitters as cyclotron resonance masers [7.9] operate using this effect. It is found that if the longitudinal motion (at a high velocity) is added to the transverse motion of particles (relative to the direction of the magnetic field), it becomes possible to design a maser based on the cyclotron autoresonance; such masers offer unique advantages. Generators of this type belong to the so-called free electron lasers (masers) discussed in chapter 11.

The negative-mass instability mentioned above limits the intensity of the beam of electrons accelerated in high-power plasma betatrons [7.10].

7.2 Acceleration of Ions in Plasma Beams with Moving Virtual Cathode

It was shown in section 7.1 that in an unstable plasma beam, ions can be accelerated transversely to the direction of beam propagation to energies much greater than that of beam electrons (this phenomenon was described there as the heating of ions). It was observed in the experiments [5.15] that longitudinal acceleration of ions (along the direction of beam propagation)

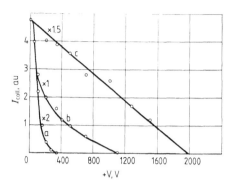

Figure 7.6   Integral energy spectra of ions moving along the beam propagation direction (curves b and c, end-face analyser probe number 1) and counter to the beam propagation direction (curve a, end-face analyser probe number 2), $V_d = 1\,kV$. Plotted on the abscissa axis is the potential of the third (control) grid of the analyser probe [5.15].

also takes place. The energies of ions accelerated along the magnetic field were measured in [5.15] by the retarding field method (see section 5.6), using multigrid probes.

The presence of longitudinal acceleration of ions follows from the comparison of retardation current–voltage characteristics of two oppositely placed end-face probes, one of which (number 1) is placed on the side of the anode and records ions moving along the beam propagation direction, while the other probe (number 2) is located on the side of the source and records ions which move in the opposite direction (the distance between the probes is 100 cm). The retardation current–voltage characteristics of the two probes are shown in figure 7.6. Figure 7.7 gives oscillograms of currents to the two probes at zero potential and at a retarding potential $V_p = +500\,V$ (the probes were within the same tube of magnetic field lines). We see that the energies of ions arriving at probe 1 are much greater than those of ions arriving at probe 2 (the particle flux to probe 1 is approximately twice as large as that to probe 2). We also find that when a retarding potential is fed to probe 2, the ion current to the probe falls off during the entire discharge pulse at an almost constant rate, which means that all ions arriving at probe 2 belong to the same group (curve a in figure 7.6); the maximum energy of these ions corresponds to the plasma potential measured by a Langmuir probe. In contrast to this, when the retarding voltage is fed to probe 1, a group of ions arriving at the probe during the first 100 ms after the discharge current has been initiated have considerably higher energies than the ions arriving during the whole remaining part of the discharge pulse. The retardation of these two groups of ions by the potential of the control grid is shown in figure 7.6 separately by curves b and c. The energies of ions accelerated along the beam are obviously greater than the beam electron energy.

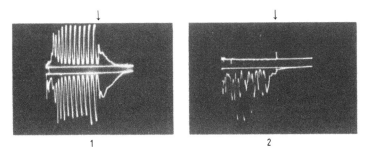

Figure 7.7   Oscillograms of ionic currents to the end-face probes number 1 (lower trace) and number 2 (upper trace) for two values of control grid potential: 1, $V_p = 0$ and 2, $V_p = 500$ V, the scanning time 1 ms, $T_d = 0.6$ ms, $V_d = 2$ kV, $I_d = 10$ A. The arrow points to the moment of turning off the plasma beam [5.15].

It is important to emphasize that ions of the group c are observed only during that interval of time in which the plasma beam travels from the cathode to the anode. These ions cease to be observed just at the moment (100 μs from the start of the discharge) when the current to the anode reaches a maximum (before that, the discharge current flows mostly perpendicularly to H, that is, towards the walls of the discharge chamber of the source — see figure 7.8). Therefore, the acceleration of the group c ions is directly related to the formation of a virtual cathode in the plasma beam, that is, the formation of a deep potential well for ions; this effect takes place always (independently of the degree of stability of the beam) at the moment of ignition of the high-power discharge, as a result of the difference between the velocities of ions and those of primary electrons (the effect is easily detected by using an inverse-current probe). Since a virtual cathode keeps oscillating in the beam during the entire remaining part of the discharge pulse, the acceleration of ions of groups b and c seems to be connected with the same mechanism of ambipolar acceleration of ions by electrons. Note that the 'diameter' of the region in which the longitudinal acceleration of ions is observed is 10–12 cm and that the total flux of these ions under the conditions of the experiments described is about 100 mA (of the order of 10% of the flux of ions which are accelerated perpendicularly to the magnetic field and end up on the walls of the vacuum chamber).

The following remarks can be made about the mechanism of the observed acceleration of ions. When a virtual cathode is formed in the electron beam, the space charge of the beam 'pushes up' the electrons travelling at its front, so that their energy ultimately grows above the initial value $W_1 = eV_0$ dictated by the accelerating potential difference V_0 in the electron beam source. Correspondingly, the potential at the virtual cathode drops to below $-V_0$, down to about $-2V_0$ [7.11–7.13]. A virtual cathode is an accelerating potential well for ions; its depth is thus about $2V_0$. If the

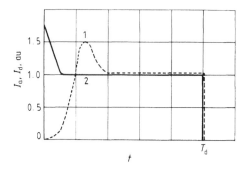

Figure 7.8   Schematic oscillograms of, 1, beam current I_a at the collector (anode) placed far from the cathode and, 2, current of electrons leaving the cathode (the discharge current I_d).

location of this well was fixed in space, the energies of accelerated ions would not exceed $2W_1 \simeq 2eV_0$. However, ion energies in a moving well may be much higher.

The phenomenon of longitudinal acceleration of ions by a moving potential well (the virtual cathode) in a shutting-off electron beam is of considerable interest both in principle and for applications, in view of the problem of novel (collective) methods of acceleration of charged particles. One such method consists in accelerating the ions to energies of several MeV (or even several tens of MeV) which takes place at the front of the intense relativistic beam propagating in the gas [7.8, 7.10–7.18]. The mechanism of acceleration of ions to energies which are several times the energy of beam electrons is definitely connected with the moving virtual cathode that builds up at the beam front: ions are accelerated only when the beam current is greater than the threshold for the formation of a virtual cathode.

The acceleration of ions in intense relativistic electron beams is thus of the same nature and is driven by the same mechanism as the phenomenon, observed in the experiments of [5.15], of acceleration of ions in beams of relatively low energy and intensity (but which is nevertheless sufficient for the formation of a (moving) virtual cathode).

Interesting and very impressive experiments on the acceleration of ions in electron beams with virtual cathode were described in [7.19]. The gas immediately ahead of the electron beam front was ionized by a laser beam propagating perpendicularly to this high-intensity electron beam; this technique made it possible to control the spatial localization of the virtual cathode in the beam. It was found possible to achieve a very efficient acceleration of ions by a moving virtual cathode, by displacing the point of intersection of the laser and the electron beams at a sufficiently high velocity ($W_1 = 1\,\text{MeV}$, $I = 30\,\text{kA}$). This approach produced an electric field up to $0.33\,\text{MV}\,\text{cm}^{-1}$ in the moving space charge of the beam over a length of 30 cm; protons were accelerated to 5 MeV, deuterons to 10 MeV,

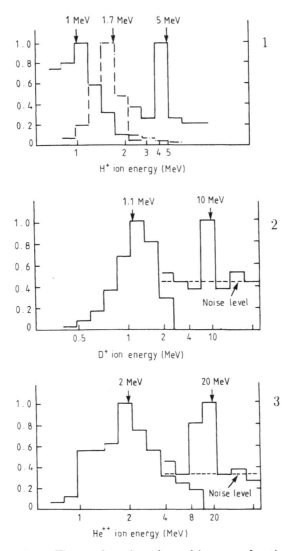

Figure 7.9 The number of accelerated ions as a function of their energy in the experiments [7.19]: 1, protons; 2, deuterons; 3, α particles; left-hand curves—no laser, dot-dash line—with the laser on, the acceleration length 10 cm; right-hand curves—with the laser on, the acceleration length 30 cm. Horizontal broken line—noise level.

alpha particles to 20 MeV and more. The role of the laser beam is decisive: in its absence, the proton energy does not exceed 1.5 MeV and that of alpha particles, about 3 MeV. The results of this work are illustrated in figures 7.9(a)–(c). The number of ions in an accelerated bunch reached 10^{12}. The feasibility of generating (under the conditions of the experiment described here) an electric field up to $1\,\mathrm{MV\,cm^{-1}}$ over a length of $100\,\mathrm{cm}$ has already been pointed out; in principle, it may be possible to create in this way fields up to $1\,\mathrm{GV\,cm^{-1}}$ over the flight length of $10\,\mathrm{m}$, that is, to accelerate protons to energies of the order of $10^3\,\mathrm{GeV}$ (and of particles with higher charge, to correspondingly higher energies). This method of acceleration appears to be very promising. The energy efficiency of this process now reaches several tens of one per cent. Paper [7.15] describes the acceleration of heavy ions (including neon, argon and krypton ions) by an intense electron beam (energy of beam particles $1.5\,\mathrm{MeV}$, current about $30\,\mathrm{kA}$) to energies of hundreds of MeV.

Note that an effective acceleration of ions to high energies was first observed in plasma beams in [7.20, 7.21]. Although the beam and plasma parameters in these experiments were not known with adequate accuracy, it can be assumed that the mechanism of ion acceleration in these experiments was close (or identical) to that described in this section.

Another mechanism of ion acceleration—by the fields of oblique Langmuir waves—was suggested in the theoretical paper [7.22].

It is difficult to refrain from citing, in conclusion, a computational work [7.23] which treats the acceleration of ions to high energies in a relativistic electron beam resulting from the Budker–Buneman instability. The most interesting result of [7.23] is the conclusion that electric fields of strength up to $20\,\mathrm{MV\,cm^{-1}}$ are generated in the beam as a result of wave collapse: this phenomenon is currently attracting considerable interest in non-linear physics of beams in plasma.

It is also of interest to point to the papers [7.24–7.27] which discuss the possibility of creating a nuclear fusion reactor based on heating the plasma by beams of high-energy ions, both light and heavy.

References

[7.1]   Baiborodov Yu T, Gott Yu V, Ioffe M S and Sobolev R I 1969 *Plasma Phys. Contr. Nucl. Fus. Res. IAEA* (Vienna) vol 2, p 213

[7.2]   Ioffe M S and Kadomtsev B B 1970 *Uspekhi Fiz. Nauk* **100** 601 (Engl. Transl. 1970 *Sov. Phys.-USPEKHI* **13** 225)

[7.3]   Nezlin M V 1969 *Plasma Phys. Contr. Nucl. Fus. Res. IAEA* (Vienna) Vol. 2 p 763

[7.4]   Neidigh R V, Alexeff I, Guest G, Jones W D, Montgomery D C, Rose D J and Stirling W L 1969 *Plasma Phys. Contr. Nucl. Fus. Res. IAEA* (Vienna) Vol. 2pp 693, 781

[7.5]   Fumelli M, Dei-Cas R, Girard P and Valckx R P G 1969 *Proc. 3rd European Conf. Contr. Fus. Plasma Phys. Symp. on Beam–Plasma Interaction* (Utrecht, The Netherlands) p 112

[7.6]   Reinmann J, Lauver M R, Patch R W, Posta S J, Snyder A and Englert G W 1975 *IEEE Trans.* **PS-3** 6

[7.7]   Roth J R 1973 *Plasma Phys.* **15** 995

[7.8]   Šunka P, Jungwirth K, Kovač I, Piffl V, Stöckel J, Ullschmied J 1977 *Proc. 8th European Conf. Contr. Fus. Plasma Phys. IAEA* (Prague) vol 1, p 108

[7.9]   Gaponov A V, Petelin M I and Yulpatov B K 1967 *Izv. VUZov, Radiofizika* **10** 1414 (Engl. Transl. 1967 *Radiophys. Quant. Electron.* **10** 794)

[7.10] Ishizuka H 1990 *Phys. Fluids* **B2** 3149

[7.11] Rostoker N 1979 *Comm. Plasma Phys. and Contr. Fusion* **5** 105

[7.12] Poukey J W and Rostoker N 1971 *Plasma Phys.* **13** 897

[7.13] Ryutov D D and Stupakov G V 1976 *Fizika Plasmy* **2** 767 (Engl. Transl. 1976 *Sov. J. Plasma Phys.* **2** 427)

[7.14] Nako F and Tajima T 1984 *Phys. Fluids* **27** 1815

[7.15] Destler W W, Floyd L E and Reiser M 1980 *Phys. Rev. Lett.* **44** 70

[7.16] Graybill S E and Uglum J R 1970 *J. Appl. Phys.* **41** 236

[7.17] Olson C L 1985 *IEEE Trans.* **NS-32** 3530

[7.18] Miller R B and Straw D C 1976 *J. Appl. Phys.* **47** 1897

[7.19] Olson C L, Frost C A, Patterson E L, Anthes J R and Poukey J W 1986 *Phys. Rev. Lett.* **56** 2260

[7.20] Plyutto A A and Kapin A T 1975 *Zh. Tekhn. Fiz.* **45** 2533 (Engl. Transl. 1976 *Sov. Phys. Tech. Phys.* **20** 1578)

[7.21] Plyutto A A, Suladze K V, Temchin S M, Mkheidze G P, Korop E D, Tskhakaya B A, Golovin I V 1973 *Zh. Tekhn. Fiz.* **43** 1627 (Engl. Transl. 1974 *Sov. Phys. Tech. Phys.* **18** 1026)

[7.22] Kondratenko A N 1987 *Pis'ma v Zh. Tekhn. Fiz.* **13** 1501

[7.23] Godfrey B B and Thode L E 1975 *IEEE Trans* **PS-3** 201

[7.24] Deutsch C 1986 *Ann. Phys.* **11** 1–111

[7.25] Arnold R C and Meyer-ter Vehn J 1987 *Rept. Progr. Phys.* **50** 559–606

[7.26] Nakai S 1990 *Nucl. Fusion* **30** 1863

[7.27] Dory R A, Houlberg W A and Attenberger S E 1989 *Comments on Plasma Phys. and Controlled Fusion* **13** 29

8 Instability of Intense Ion Beam with Neutralized Space Charge in Magnetic Field

8.1 About this Chapter

Intense beams of fast ions in strong magnetic fields are now used in a considerable number of phyasical, experimental and stationary set-ups; for example, electromagnetic isotope separators and magnetic traps for plasma with external injection of fast ions. The efficiency of using ion beams in such systems is determined to a large extent by phenomena connected with the neutralization of the beam space charge [8.1]. At the same time, there are several common features in the design of these set-ups, both in the method of creating the ion beam (the ionization of the working gas or vapour by electron impact and the extraction of ions from the discharge by a strong electric field) and in the method of beam transport in a strong magnetic field. We can assume, therefore, that phenomena determining the neutralization of the space charge of the ion beam in these systems and its violation (de-compensation) are likely to be connected by a common mechanism.

Let us consider this mechanism under specific conditions of operation of an electromagnetic isotope separator with 180° beam deflection in transverse magnetic field (figure 8.1) [8.1–8.3]. As we shall see from the experimental data obtained and from their interpretation, the mechanism is fairly universal; it is independent of whether the magnetic field is transverse or longitudinal with respect to the beam, and is also independent of the type of ion source. Limitations of the universality of this mechanism (which are of theoretical rather than practical significance) stem from the following two conditions: (1) electrons neutralizing the space charge of ions in the beam are magnetized while ions are not, and (2) the beam is created by a plasma source. It is not difficult to see that these limitations are in fact merely formal. The results outlined below should therefore be kept in mind

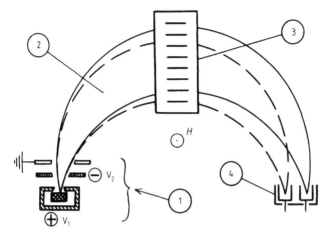

Figure 8.1 Schematic diagram of electromagnetic separation of isotopes with 180° beam deflection by transverse magnetic field: 1, ion source; 2, fast (primary) ion beam; 3, secondary ion energy analyser; 4, beam receptor.

by all experimentalists who work with intense ion beams.

Two types of plasma are found in any electromagnetic separator. First, there is a quasineutral beam of fast ions with energies of tens of keV, which is used directly in the process of separation. The space charge of the beam is neutralized by electrons that are produced by the beam in the residual gas. Second, there is the auxiliary gas discharge in the ion source, sustained in a strong longitudinal magnetic field; the gas discharge plasma is created by ionizing the vapour of the working substance by an electron beam with an energy of several hundreds of electron volts. The ion beam is produced by extracting ions from the discharge by a strong electric field; it then propagates perpendicularly to the magnetic field lines. In both cases, we deal with the beam plasma subject to instabilities. These instabilities violate the neutralization of the space charge of the beam, that is, they de-focus the beam and spoil the process of isotope separation [5.30, 5.31, 8.1, 8.2]

An experimental investigation of these instabilities was conducted in [5.30, 5.31], where efficient methods of controlling them were also found. An application of these methods made it possible to achieve a high degree of neutralization of the space charge of the ion beam at any current provided by the ion source in the isotope separation set-up.

Most of the experimental data in [5.30, 5.31] were obtained with an electromagnetic mass separator with 180° beam deflection (see figure 8.1). The beam ion energy was 30 keV, magnetic field strength 2×10^3 Oe, ion trajectory length 300 cm, beam height 10–20 cm, at residual gas pressure of about 1×10^{-5} mm Hg.

The need to neutralize the space charge of the ion beam for the high-output separation of isotopes was obvious already in the first steps of the electromagnetic method [8.1, 8.2]. Therefore, an analysis of factors capable of preventing this neutralization was started already in the earliest papers on the development of the method. Without going into a detailed history of this research (it can be found in [5.31]), I will mention the main results, rephrased in terms of the current understanding of the phenomena.

The root cause of the violation of space charge compensation in the ion beam are the discharge oscillations in the ion source. An even more important factor producing this phenomenon is a very efficient mechanism of oscillation amplification in the ion beam. The oscillation amplification coefficient K is found to be an abrupt and non-monotone function of the ion beam current: as the current increases from relatively low values, K first grows rapidly but then reaches a maximum and starts to fall off. Therefore, if some discharge parameter in the ion source is varied and this changes the ion beam current, the main effect on the state of neutralization of the ion beam space charge is determined not so much by changes of discharge oscillations as by changes in the oscillation amplification coefficient in the ion beam. (E.g., as the vapour pressure of the working substance is increased, the discharge oscillation amplitude decreases; however, the amplitude of oscillations of the space charge in the ion beam increases as a result of beam current increase and the increased coefficient K, etc.)

The work [5.30, 5.31] created a theoretical and practical basis for solving the problem of neutralization of ion beams regardless of the current produced by the available ion sources for the electromagnetic isotope separation.

The measurements in [5.30, 5.31] were mostly carried out with beams of medium-mass ions: Ca^+ ($M = 40$), Zn^+ ($M = 64$), Cd^+ ($M = 114$), Te^+ ($M = 128$). In a more recent paper [8.4], a similar study was carried out with a beam of considerably heavier lead ions Pb^+ ($M = 206$). The results of [5.30, 5.31] and [8.4] are physically equivalent. It was also shown in [8.4] that as the current density in the beam changes, a simple scaling relation governs the dependence of the potential and of the angular divergence of beams of ions of different masses on this density. These dependences characterize the action of a universal amplification mechanism discovered in [5.30, 5.31]. This mechanism is discussed in the next section. (As for the mechanism of oscillations in the ion source discharge, we refer the reader to a detailed publication [5.31].)

8.2   Mechanism of Oscillation Amplification in Ion Beams

The instabilities of the gas discharge plasma in the ion source cause oscillations in the ion beam current I. However, the immediate cause of

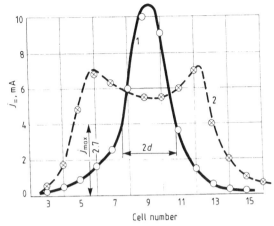

Figure 8.2 Current density distribution across the Zn^+ ion beam near the initial point of its trajectory ($y = 15$ cm), for two values of current [5.30, 5.31]. 1, $I = 100$ mA; 2, $I = 150$ mA.

violation of space charge neutralization in the beam are the oscillations of the ion current density j. Therefore, to understand the mechanism of ion beam decompensation, it is necessary to find the relation between the oscillations of I and those of j, or, to put it differently, to find the relation of the distribution of the current density transversely to the beam axis, $j(x)$, to the beam current I. (Unless specially explained otherwise, the current I invariably stands for the DC component $I_=$.)

This distribution found in [5.30, 5.31] is shown in figure 8.2, where the letter d stands for the beam 'half-width', that is, for its transverse dimension corresponding to an e-fold decrease in current density. The curve $d(I)$ is shown in figure 8.3(a). I will discuss its interpretation later; the following important points must now be emphasized.

1. The very fact that d is a function of I signifies that the relation between I and j is non-linear. Therefore, the relative amplitude of current density oscillations in the ion beam, $j_\sim/j_=$ (i.e., the ratio of the oscillation amplitude to the DC component of the current density) does not equal the relative oscillation amplitude of the beam current, $I_\sim/I_=$. Hence, a process of oscillation amplification is at work in the beam, and the amplification coefficient $K(I) = (j_\sim/j_=)/(I_\sim/I_=)$ is not equal to unity.

2. The function $d(I)$ is non-monotone and has an abrupt dip at some value of the current ($I = I_{min}$). (This type of function $d(I)$ is typical of all systems with a high space charge in the acceleration path [5.31, 8.5].) It is natural to expect, therefore, that the value of $K(I)$ in the neighbourhood of the point $I = I_{min}$ has a maximum, since it is precisely at $I = I_{min}$ that the relative change in the beam half-width (in response to changes in the current) is the greatest. In order to show it, we will find the expression for

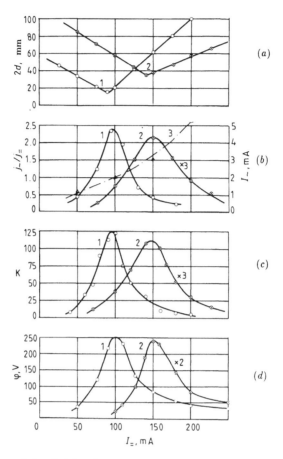

Figure 8.3 (*a*) ion beam half-width *d*; (*b*) relative current density oscillation amplitude $j_\sim/j_=$ (solid curves) and the current oscillation amplitude $I_\sim$; (*c*) oscillations amplification coefficient K; (*d*) beam potential φ as functions of beam current $I_=$ for the values of the intermediate electrode potential V_2 (see figure 8.4): 1, $V_2 = -5\,\text{kV}$; 2, $V_2 = -15\,\text{kV}$. Figures (*a*) and (*b*) refer to the initial part of the trajectory ($y = 15\,\text{cm}$), and figures (*c*) and (*d*) refer to its middle section. A Zn^+ ion beam, $V_1 = 30\,\text{kV}$, $H = 2.1 \times 10^3$ Oe, $p = 1 \times 10^{-5}$ mm Hg [5.30, 5.31].

$K(I)$. We will need an analytic form of the function $j(x, I)$, for instance

$$j(I, x) = j(I, 0) \exp(-x^2/d^2) \qquad (8.1)$$

(here $j(x, I)$ is assumed to be the linear current density) or, since

$$I = \int_{-\infty}^{+\infty} j(I, x)\,\mathrm{d}x = j(I, 0)\sqrt{\pi}\,d$$

we have

$$j(I,x) = \frac{I}{\sqrt{\pi}\, d} \exp(-x^2/d^2). \tag{8.2}$$

As follows from figure 8.3(a), $d \simeq d_{min} + \beta(I - I_{min})$, where d_{min} is the minimal beam half-width; note that on crossing the point $I = I_{min}$, the parameter β both changes the value and reverses the sign (indeed, $\beta > 0$ if $I > I_{min}$ and $\beta < 0$ if $I < I_{min}$). Equation (8.2) then implies

$$\frac{\mathrm{d}j}{\mathrm{d}I} = \frac{I\beta(2x^2 - d^2) + d^3}{\sqrt{\pi}\, d^4} \exp(-x^2/d^2). \tag{8.3}$$

From (8.3) and (8.2), we find the oscillation amplification coefficient

$$K(x,I) \equiv \frac{\mathrm{d}j/j}{\mathrm{d}I/I} = 1 + \frac{\beta I(2x^2 - d^2)}{[d_{min} + \beta(I - I_{min})]^3}. \tag{8.4}$$

Let us consider the main consequences of (8.4).

1. The current density oscillations in the inner and outer regions of the beam are shifted in phase by 180°: the oscillations of j and I are in phase if $|x| > d/\sqrt{2}$, and they are in antiphase if $|x| < d/\sqrt{2}$. If $|x| = d/\sqrt{2}$, then $K(x,I) = 0$. If $I > I_{min}$, this means that a fast increase in I produces an excess of ions at $|x| > d/\sqrt{2}$ and a deficiency of ions at $|x| < d/\sqrt{2}$; in the meantime, the beam space charge undergoes decompensation in the outer region while quasineutrality is maintained in the inner region (owing to the rapid removal of 'excess' electrons along the magnetic field to the lids of the separation chamber). Conversely, if the beam current rapidly decreases, the inner region is decompensated while the negative charge excess in the outer region vanishes. A part of the beam (outer or inner) is thus always decompensated in both phases of oscillations. Hence, the beam has a considerable positive potential with respect to (grounded) chamber lids. (Similar phenomena occur at $I < I_{min}$ but since $\beta < 0$, the inner and outer regions of the beam reverse their roles in the argument.)

2. In the entire beam, except at the points $x = \pm d/\sqrt{2}$, we have

$$K(x,I) \simeq 1 + \frac{|\beta|I}{d} \simeq 1 + \frac{|\beta|I}{d_{min} + \beta(I - I_{min})}. \tag{8.5}$$

We see that the oscillation amplification coefficient is a non-monotone function of current and passes through a maximum at $I = I_{min}$:

$$K(I)_{max} \simeq 1 + |\beta|I_{min}/d_{min}. \tag{8.6}$$

In the vicinity of the ion source, $|\beta|I_{min}/d_{min} \simeq 5$ (see figure 8.2) and $K_{max} \simeq 6$.

3. The oscillation amplification coefficient increases along the beam trajectory. This occurs because the beam decompensation in the initial part of the trajectory causes additional beam divergence at points farther along the beam, so that the degree of decompensation of the space charge also gradually increases; this is equivalent to increasing the coefficient β in (8.6). We see, in view of this factor, that the value of $K(I_{max})$ far from the ion source, say, midway along the beam, must be as large as several tens or even larger. An elementary evaluation will show (see section 5.1) that the rate of variation of the space charge in the ion beam is much greater than the rate of generation of electrons, so that strong electric fields arise in the beam. As an indicator of these fields, we are going to use below the maximum potential in the beam in the cross section at the midpoint of its trajectory. This potential is proportional to the oscillation amplitude of the beam current density [5.30]. (The methodology and the results of measuring the potential and the electric fields in beams are described in section 5.6.4.)

4. As follows from (8.3) and (8.4), there is no correlation between oscillations in the discharge (they determine the quantity dI/dt) and oscillations of current density in the ion beam (dj/dt).

5. The factor responsible for the amplification mechanism discussed above is the dependence $d(I)$ itself, which results in a non-linear relation between I and j. The following explanation can be given to this dependence [5.30, 5.31] (see figure 8.3(a)). The gas discharge plasma, acting as the ion emitter, has the shape of a concave meniscus at the interface with the optical section of the source; the meniscus curvature is greater, the lower the plasma density and the higher the electric field strength in the optical section. The beam divergence in the XY plane perpendicular to H is, therefore, relatively large at low beam current (i.e, at low plasma density; see lines 1–1 in figure 8.4). As the current increases (this is done by increasing the plasma density) the beam divergence first falls off. There are two reasons for this: a decrease in meniscus curvature and an increase in beam repulsion in the optical section (line 2–2 in figure 8.4). However, it is not difficult to see that beginning with a certain current $I = I_{min}$, these factors bring about not a decrease but an increase in the beam divergence (lines 3–3 in figure 8.4). The current I_{min} corresponding to the minimum half-width d is determined by factors that affect the meniscus curvature and the mutual repulsion of the space charge of ions in the optical section of the source. Thus I_{min} increases as the potential of the intermediate electrode increases (II in figure 8.4), as the slit width in the electrode is reduced, and as the ion mass M gets smaller. For example, a decrease in M reduces the beam space charge proportionally to $\sqrt{M}$, so that the value of I_{min} must, other conditions being equal, increase proportionally to $1/\sqrt{M}$.

Let us look now at a comparison of the experimental data with the above point of view on the mechanism of amplification of oscillations in the ion

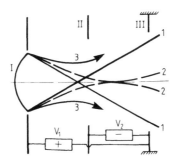

Figure 8.4 Horizontal divergence in the XY plane of the ion beam in the accelerating (optical) system of the source: I, plasma meniscus in the discharge chamber; II, intermediate electrode; III, the third electrode at the potential of the separation chamber walls; 1–1, 2–2 and 3–3 are the trajectories of ions. The magnetic field and the height of the working slit in the discharge chamber are perpendicular to the plane of the figure. V_1 is the accelerating voltage, $W_1 = eV_1$ is the beam ion energy and V_2 is the potential of the intermediate electrode.

beam.

1. *The observations of the oscillation amplification process and of the growth of the amplification coefficient along the ion beam.* Figure 8.5 plots the distribution of the constant and of the alternating components of the current density, reproduced from [5.30, 5.31], across the ion beam in two cross sections: (a) close to the ion source ($y = 15\,\text{cm}$) and (b) at the midpoint of the beam trajectory ($y = 140\,\text{cm}$). We see that while $I_\sim/I_= = 2 \times 10^{-2}$,

$$j_\sim/j_= = \begin{cases} 6 \times 10^{-2} & \text{in case } (a) \\ 80 \times 10^{-2} & \text{in case } (b) \end{cases}$$

that is,

$$K(x = 0) = \begin{cases} 3 & \text{close to the ion beam} \\ 40 & \text{at the midpoint of the beam trajectory.} \end{cases}$$

It is important to note that the frequency spectrum of oscillations in the beam is the same as in the source discharge, that is, we discuss the amplification of oscillations in the discharge, not the generation of some new oscillations.

2. *The correspondence of the maximum of the oscillation amplification coefficient to the minimum of angular divergence and to the maximum of the ion beam potential.* Figure 8.3(a)–(d) shows the functions $d(I)$, $K(I)$, $f(I) = j_\sim/j_=$ and $\varphi(I)$ (φ is the beam potential). We see that the behaviour of these functions agrees quite well with the standpoint outlined above.

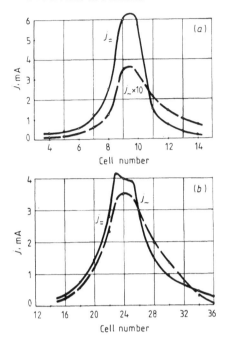

Figure 8.5   The distribution of the constant $(j_=)$ and the alternating $(j_\sim)$ components of the linear current density across the Zn^+ ion beam at two distances y from the source at $V_1 = 30\,kV$ and $I_\sim/I_= = 2 \times 10^{-2}$ [5.30, 5.31]. a, $y = 15$ cm, b, $y = 140$ cm.

3. *The shift of the oscillation amplification maximum in response to a change in ion mass and electric field strength in the optical system of the ion source.* Figure 8.3 has been plotted for two values of the potential of the intermediate electrode: $-5\,kV$ and $-15\,kV$. We see that in accordance with the point of view under discussion here, the extremums of the functions $d(I)$, $\varphi(I)$ and $K(I)$ are 'translated' in the 'correct' direction along the axis of the ion current.

Figure 8.6 gives the dependence $\varphi(I)$ for beams of ions of different mass, beginning with $M = 7$ (lithium) and ending with $M = 114$ (cadmium). We see, first, that the lower M, the farther the maximum is shifted towards larger I, and second, that the translation is approximately proportional to $\sqrt{M}$. These facts also confirm the qualitative theory presented above.

4. *The absence of direct correlation between the oscillations of beam current density and those of beam current.* Figure 8.3(b) also shows how the amplitude $I_\sim$ of beam current oscillations (proportional to the oscillation amplitude in the ion source discharge) depends on the beam current $I_=$. Indeed, there is no correlation between the variations of the oscillation amplitude of beam current density and the discharge oscillations. This is

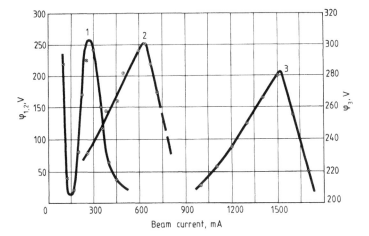

Figure 8.6 Beam potential as a function of current for beams of ions of different mass: 1, $M = 114$ (Cd^+); 2, $M = 40$ (Ca^+); 3, $M = 7$ (Li^+) [5.31].

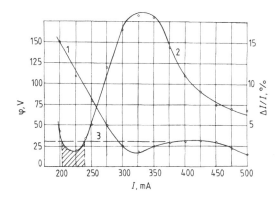

Figure 8.7 The relative amplitude of current oscillations, 1, and potential of a Zn^+ ion beam, 2, as functions of beam current. Horizontal line, 3, shows the admissible beam potential, φ_{adm} at which the beam focusing is still acceptably good [5.30, 5.31]. The arrow points to the shadowed region of the working range ($\varphi \leqslant \varphi_{adm}$). $V_1 = 30\,kV$, $V_2 = -25\,kV$.

only natural since

$$j_\sim/j_= = K(I)\, I_\sim/I_= \qquad (8.7)$$

and since the relation between $I_\sim$ and $j_\sim$ is essentially non-linear; moreover, figures 8.3 and 8.7 demonstrate that the form of the function $j_\sim(I)$ is dictated above by the form of $K(I)$.

5. *The amplification coefficient and the oscillation amplitude.* We see from figures 8.3 and 8.5 that the value of K at the midpoint of the beam trajectory reaches several tens or even hundreds; in this last case, $j_\sim(I)$

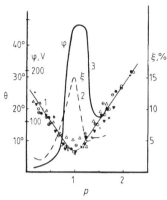

Figure 8.8   Ion beam potential φ and divergence angle Θ as functions of the similarity parameter P which is controlled by changing ion mass M and voltage $V_0 = V_1 + |V_2|$. $\square$, $M = 207$, $V_0 = 20\,\text{kV}$; $\blacklozenge$, $M = 207$, $V_0 = 25\,\text{kV}$; $\bullet$, $M = 207$, $V_0 = 30\,\text{kV}$; $\circ$, $M = 207$, $V_0 = 35\,\text{kV}$; $\blacktriangle$, $M = 207$, $V_0 = 40\,\text{kV}$; $+$, $M = 207$, $V_0 = 45\,\text{kV}$; $\triangle$, $M = 64$ (zinc), $V_0 = 35\,\text{kV}$; $\ominus$, $M = 23$ (sodium), $V_0 = 35\,\text{kV}$ [8.4].

is considerably greater than j_-. As was already mentioned in [5.30, 5.31], the rate of variation of the space charge density in the beam, $\omega j_\sim/uh$, is considerably greater than the rate of accumulation of the electron space charge, $(j_=/h)n_0\sigma_i u$ (u is the velocity of ions, $\omega/2\pi$ is the oscillation frequency, h is the beam height, n_0 is the residual gas density and σ_i is the electron production cross section). For instance, if $p \simeq 1 \times 10^{-5}\,\text{mm}$ Hg (residual gas), the ratio of these quantities reaches 10–20, or even higher. In principle, this situation remains unchanged if the ionization of the gas by neutralizating electrons is taken into account.

8.3   Similarity Laws for Plasma Ion Sources

Figures 8.8 and 8.9 show the divergence angle and beam potential measured in [8.4] in beams of different mass ions (atomic mass 23 (Na), 65 (Zn) and 207 (Pb)) as functions of the so-called similarity parameter

$$P = \frac{9\pi\delta^2}{0.4V_0^2} \frac{j}{\sqrt{2e/M}} \qquad (8.8)$$

where δ is the distance between the accelerating electrodes of the ion source, V_0 is the total accelerating voltage (the potential difference between electrodes I and II in figure 8.4), j is the ion beam current density and M is the atomic mass of an ion.

The parameter in figures 8.8 and 8.9 is the ratio $\psi = V_1/V_0$, where V_1 is the potential of electrode I (in figure 8.4) with respect to the ground.

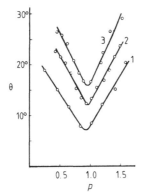

Figure 8.9   Angle of divergence of the ion beam as a function
of similarity parameter P for $M = 23$ (sodium) and $V_0 = 35\,\text{kV}$
[8.4]. 1, $\psi = 0.14$; 2, $\psi = 0.23$; 3 $\psi = 0.40$.

We see that both curves, obtained by varying the parameters of the ion
beams over a very wide range ($j = 6 - 140\,\text{mA/cm}^2$, $V_0 = 20 - 45\,\text{kV}$,
$M = 23 - 207$), do obey a simple similarity law (see also [8.5]). This fact
points to the universality of the suggested mechanism of amplification of
space charge oscillations in ion beams.

8.4   Methods for Drastically Improving the Neutralization of the Ion Beam Space Charge

I will now describe, without going into the methods of reducing the
amplitude of oscillations in the ion source discharge (see detailed papers
[5.30, 5.31]), the basic methods of improving the neutralization of ion
beams; these methods make use of the above oscillation amplification
mechanism in an ion beam.

   It was shown above that the maximum of the oscillation amplification
coefficient can be shifted along the ion current scale by varying the potential
V_2 of the intermediate electrode of the accelerating system of the ion source
and also by changing the dimensions of this system. This effect is shown in
greater detail in figure 8.10. Technical specifics determine what precisely
has to be undertaken in practical conditions. For instance, if the ion source
works with acceptable stability at high beam current, it is advisable to
reduce $|V_2|$, thereby shifting the maximum of $K(I)$ to the left and achieving
efficient neutralization of the space charge in the beam at high ion currents,
far beyond the maximum of $K(I)$. Otherwise, one needs to increase $|V_2|$;
this improves the neutralization of the beam charge at moderately high ion
currents in the range to the left of $K(I)$. Obviously, the former method is
more attractive and is used more frequently.

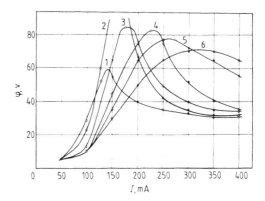

Figure 8.10   Shift of the $\varphi(I_=)$ curve in response to the variation of the intermediate electrode potential; Zn^+ ion beam, the height of slit in the discharge chamber 10 cm [5.30, 5.31]. 1, $|V2| = 0$; 2, 5 kV; 3, 10 kV; 4, 15 kV; 5, 20 kV; 6, 30 kV.

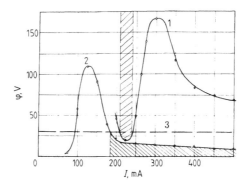

Figure 8.11   Effect of source geometry and potential of the intermediate electrode on the neutralization of a Zn^+ ion beam: 1, $h = 18$ cm, $\delta_1 = \delta_2 = 3$ mm, $\delta_3 = 7$ mm, $V_2 = -25$ kV; 2, $h = 14$ cm, $\delta_1 = 2$ mm, $\delta_2 = 4.5$ mm, $\delta_3 = 9$ mm, $V_2 = -15$ kV [5.30, 5.31]; 3, $\phi = \phi$ adm. The shadowed region shows the working zone with admissible beam potential $(\varphi = \varphi_{adm})$.

The combined effect of the methods of suppressing oscillations in the discharge (see [5.31, 5.32]) and of suppressing their amplification is shown in figure 8.11, which uses a beam of Zn^+ ions as an example. The figure also shows the admissible value of the potential (φ_{adm}) at which the beam focusing is still acceptably good. The remaining notation in the figure: h is the height of the working slit in the gas discharge chamber of the ion source (i.e., its size along H); δ_1 is the width of this slit, δ_2 is the distance from the discharge filament to the outer plane of electrode I (see figure 8.4) and δ_3 is the width of the slit in electrode II (see figure 8.4). It is clear that

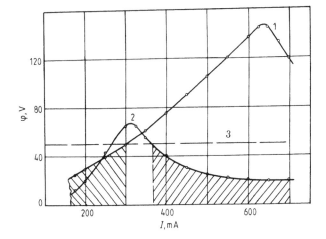

Figure 8.12   Same as in figure 8.11 but for a Ca$^+$ beam (for details, see [5.31] and [1.17]).

the procedures outlined above make it possible to substantially raise the current in well-neutralized ion beams. The beam intensity is then limited by the 'ceiling' of the ion source itself, which is dictated by the Langmuir law $\frac{3}{2}$ and by the increasing angular divergence of the beam. A similar result for a beam of Ca$^+$ ions is shown in figure 8.12 (for details, see [5.31] and [1.17]).

The methods suggested to improve space charge compensation in ion beams are sufficiently universal (see also [8.4]).

The phenomena described in this chapter must also be of great interest in systems of injection of heavy particle flows into nuclear fusion plasma traps [8.6, 5.17, 5.49].

References

[8.1]   Smith L P, Parkins W E and Forrester A T 1948 *Phys. Rev.* **72** 989

[8.2]   Koch J, Dawton R *et al* 1958 *Electromagnetic Isotope Separators and Applications of Electromagnetically Enriched Isotopes* (New York: Interscience)

[8.3]   Love L O 1973 *Science* **182** 343

[8.4]   Raiko B I 1963 *Zh. Tekhn. Fiz.* **33** 244 (Engl. Transl. 1963 *Sov. Phys. Tech. Phys.* **8** 175)

[8.5]   Green T S 1974 *Reports on Progr. Phys.* **37** 1257

[8.6]   Artsimovich L A 1963 *Controlled Thermonuclear Fusion Reactions* (Moscow: Fizmatgiz) Section 2.5.

9 Electric Double Layers in Electron, Ion and Plasma Beams and the Mechanism of Aurora Borealis

9.1 Double Layers in the Plasma (Theory)

Interest in electric double layers reported in Langmuir's classical papers [5.14] and also studied by Bohm [5.13] has been revived in recent years in plasma physics [9.1–9.12]. A double layer can also be treated as a plasma capacitor in which the quasineutrality of the plasma is substantially violated and the potential difference in the layer ($\Delta\varphi$), caused by the separation of charges, may be considerably higher than the electron (and ion) temperatures. The double layers we are to discuss in this chapter are precisely such strong double layers†. Double layers play a decisive role not only in laboratory plasma but also in natural phenomena, such as the *aurora borealis* [9.5–9.8].

Mechanisms of formation of double layers in the plasma may be quite different. One of such, with which we start the discussion, determines the ignition of the discharge with a hot cathode in a strong longitudinal magnetic field (see section 5.3). The ignition of a discharge is defined as a state in which the entire discharge voltage ($V_d = W_1/e \gg T_e/e$) applied between the beam source (the cathode) and the beam collector (the anode) is concentrated in a thin cathode layer, so that the beam current increases sharply, obeying the familiar Langmuir $\frac{3}{2}$ law

$$j_1 \simeq \frac{2}{9\pi} \left(\frac{2e}{m}\right)^{1/2} \frac{V_d^{3/2}}{d^2} \tag{9.1}$$

---

† For details on weak double layers, I refer the reader to the extensive literature available [9.1, 9.3, 9.4, 9.8].

where j_1 is the beam density and d is the layer thickness (the coefficient 2 appears because of the effect of the positive charge of plasma ions). According to Langmuir [5.14], the condition of stability of the double layer (i.e. the condition of stability of the discharge) requires that there exist a special relation between the beam electron density j_1 and the density of the ion current (i.e. countercurrent) from the plasma, j_+, namely:

$$j_1 \leqslant \gamma (M/m)^{1/2} j_+ \tag{9.2}$$

where $j_1 = .en_1 u$ and γ is a numerical coefficient which slightly depends on the properties of the cathode surface (its typical value is $\gamma = 3/2$). According to Bohm's theorem [5.13], a double layer is stable only if plasma ions (driven by the electric field of the double layer which penetrates into the plasma) arrive at the layer boundary with velocities

$$v_+ \geqslant (T_e/M)^{1/2}. \tag{9.3}$$

These velocities correspond to the energy of directed motion of at least $T_e/2$. Correspondingly, the plasma density at the boundary of the layer, (n_+), is, by virtue of Boltzmann's law, less than the density n_2 of non-perturbed plasma by a factor of $\sqrt{2.7}$; the ion current density in the layer is $j_+ = en_+ v_+$ (it is assumed that the ion temperature $T_+ \ll T_e$). Taking Bohm's theorem into account, we find the following condition of stability of the double layer:

$$n_1 u \lesssim \gamma\sqrt{2}\,\frac{n_2 v_2}{4} \tag{9.4}$$

where $v_2 = (8T_e/\pi m)^{1/2}$ is the average thermal velocity of plasma electrons. Since $\gamma\sqrt{2} \simeq 1$, equation (9.4) implies that

$$n_1 u \leqslant \tfrac{1}{4} n_2 v_2 \qquad \text{or} \qquad j_1 \leqslant j_2 \tag{9.5}$$

where $j_2 = \tfrac{1}{4} en_2 v_2$ is the density of the random current of plasma electrons. Therefore, the condition of existence of a stationary double layer at the cathode or (which is an equivalent statement) the condition of existence of an electron beam in a discharge with a hot cathode, is that the beam current be not higher than the random current of plasma electrons. If this condition is satisfied, a beam–plasma system is formed and the plasma density at a sufficiently high beam electron energy ($W_1 = mu^2/2 \gg T_e$) is found to be much greater than the beam density:

$$\frac{n_2}{n_1} \geqslant 4\,\frac{u}{v_2} \gg 1. \tag{9.6}$$

In the particular case discussed here, the double layer is stationary and 'pressed on' to the cathode. In principle, the Langmuir–Bohm conditions could be met far from the electrodes, for instance, midway across the discharge gap. In this situation (with a stationary layer) another condition, symmetric to (9.3), must be satisfied by the velocities with which electrons arrive at the layer boundary:

$$v_1 \geqslant (T_+/m)^{1/2}. \tag{9.7}$$

A situation often arises in real plasma when conditions (9.2)–(9.7) are not satisfied for a *stationary* layer. In such cases, it is possible that the same conditions hold for a *moving* layer, that is, they hold in a frame of reference which is stationary with respect to a layer moving in the laboratory reference frame. If this scenario does take place, a stable moving double layer is produced.

We have discussed the simplest situation, in which only two groups of charged particles participated in the formation of the double layer: free electrons and free ions. The actual situations are more complex since at least two more groups of particles must generally be taken into account in a layer separated from the electrodes: slow 'trapped' electrons which penetrate into the layer only a short distance on the 'ion' side and are confined by the positive potential of the layer, and slow 'trapped' ions which penetrate into the layer a short distance on the 'electron' side and are confined by the negative potential of the layer. These phenomena are discussed in papers [9.9–9.11].

9.2   Double Layers in the Laboratory. A Virtual Cathode in an Electron Beam as a Double Layer

A large number of experiments are known at present in which strong double layers were observed, with a potential drop reaching several thousand times the electron temperature (see, e.g., [9.12–9.18]). As a rule, these are moving layers. Typically, their formation implies restrictions on the current flowing through the system. The conditions of formation of these double layers correspond to the Langmuir–Bohm conditions (with the motion of the layers taken into account).

The layers arise both in the case when an external voltage is applied to the system and when the external voltage is zero. In the latter case, the formation of double layers is connected, among other things, with the Pierce and Budker–Buneman instabilities [9.19–9.22] and also with the dynamics of the arising virtual cathode (at sufficiently high currents) [9.9–9.11].

One of the types of double layer in a plasma is the virtual cathode which arises in the plasma beam (Chapter 6) if

$$n_1 u > \tfrac{1}{4} n_2 v_2. \tag{9.8}$$

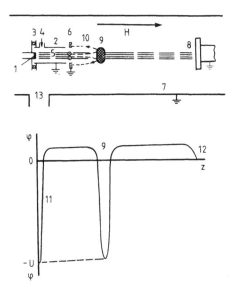

Figure 9.1 Distribution of potential φ along the electron beam with virtual cathode: 1, cathode 1 cm in diameter at a negative potential U; 2, discharge chamber of the beam and plasma source (2 cm in diameter, 15 cm long, at zero potential); 3, insulator; 4, gas injection; 5, beam; 6, 'inverse current probe' (indicator of virtual cathode); 7, vacuum chamber; 8, anode; 9, virtual cathode; 10, trajectories of the majority of beam electrons after the formation of the virtual cathode; 11 and 12, positive space charge layers at the cathode and the anode; 13, pumping-out pipe [6.12].

At first glance, this condition contradicts condition (5.21) of ignition of discharge, that is, the condition of existence of the plasma beam. This contradiction is only an apparent one: the point is that the plasma density at the cathode, where condition (5.21) is satisfied, is higher than when far from the cathode where conditions (6.31) and (9.8) are met. These (seemingly contradictory) conditions are brought into agreement in the experimental situation by producing a discharge with longitudinal plasma density gradient (correspondingly, with a gradient of the neutral gas density).

Therefore, if the beam current is not higher than the random current of plasma electrons, only one double layer exists in the discharge with hot cathode, namely, the layer at the cathode boundary of the beam. If, however, the beam current is greater than the random current of plasma electrons, then two double layers appear in the system: one close to the cathode and the other in the region of the virtual cathode. The second layer consists of two contiguous layers: there is an excess of negative charge at the minimum of the potential, and there is an excess of positive charge closer to the boundaries of the plasma (in front of and behind the virtual cathode). The potential distribution along the beam length in the two-layers mode is shown schematically in figure 9.1, together with the schematic diagram of the experimental set-up [6.12]. A proper term can be suggested for the region of the virtual cathode (9 in figure 9.1): the 'triple layer'. The thickness of the virtual cathode layer is determined by a relation which is essentially similar to (5.11); within an order of magnitude, this thickness is about a fraction of one centimetre [6.12]. Note that generally there is another double layer under the conditions described: that at the anode (12 in figure 9.1); the potential drop across it is several T_e/e, and it is necessary to maintain the quasineutrality of the plasma column.

Double layers connected with the formation of a virtual cathode in beam–plasma systems are formed owing to two instabilities analysed above: the beam drift and the Pierce instabilities; as for the Buneman instability, it was mentioned above that it does not produce a virtual cathode under the experimental conditions outlined. The main role in the formation of the internal double layer in the discharge with hot cathode (far from the plasma boundary) was played by the beam–drift instability; in other experimental conditions (in the absence of the magnetic field [9.19, 9.21]) this was the Pierce instability (in this connection, see also [6.12, 9.22]).

Interesting manifestation of chaos in systems with electric double layers in the plasma, including the evolution via period doubling and intermittency, were reported in [9.23] and [9.24].

9.3   Double Electric Layer ('Virtual Anode') in an Ion Beam Propagating in a Plasma, in Earth's Magnetic Dipole Model. The Mechanism of Polar Auroras

The problem of electric double layers in a plasma (one of the versions of such layers, connected with the virtual cathode of the electron beam was discussed in the preceding section) is now an object of intense experimental and theoretical investigation. This attention was stimulated by the fact that electric double layers are a candidate for a very probable mechanism of the discrete aurora borealis, which is an immensely intriguing natural phenomenon. A point of view, supported by theory, by numerical simulations and by physical experiments, makes use of this mechanism; it is shared by a number of researchers. This approach will be presented below.

According to a number of rocket (and space probe) observations of the last decade, an electrostatic field directed from the ionosphere to the magnetosphere arises under certain conditions in the plasma surrounding the Earth, at distances of the order of one Earth radius; this field forms the so-called 'inverted V', or 'lambda', structures [9.5–9.8, 9.25–9.32] an example of which is illustrated schematically in figure 9.2 (on the magnetic

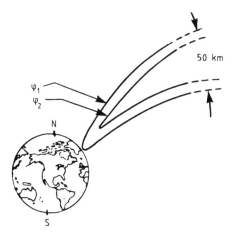

Figure 9.2 Model of electric double layer (inverted V structure) which accelerates electrons towards the Earth. Potential $\varphi_1 = +(3 - 10)\,\text{kV}$, $\varphi_2 < \varphi_1$.

configuration of the extraterrestrial plasma, see [9.34]). The strength of this electric field and its spatial extension are such that the total potential difference (the positive polarity being on the side of the ionosphere) reaches 3 to 10 kV.

It is owing to the potential difference across which electrons pass in the inverted V structures (when they arrive from the magnetospheric tail), that intense fluxes of electrons become capable of reaching the ionosphere (overcoming the strong magnetic mirror field) and get accelerated sufficiently to excite gas molecules in the atmosphere and to generate (discrete) aurora borealis. The need to accelerate the magnetospheric electrons in an adequate potential difference follows from formula (9.9) given below; if the potential difference is smaller, the flux of electrons reaching the ionosphere is too low to produce the effect we call *aurora borealis*. This formula [9.34, 9.35] shows the density of the electron current through a magnetic mirror as a function of the mirror ratio B_1/B_2 and of the accelerating potential difference $\varphi_\parallel$ (we assume a Maxwellian energy distribution function at a temperature T_e for electrons moving far from the mirror):

$$j_\parallel = \frac{env}{4} \frac{B_1}{B_2} \left[1 - \left(1 - \frac{B_2}{B_1} \right) \exp\left(-\frac{e\varphi_\parallel}{T_e[B_1/B_2 - 1]} \right) \right] \qquad (9.9)$$

where B_1 is the magnetic field in the ionosphere and B_2 is the field at the initial point from which an electron must reach the ionosphere, moving downwards and having overcome the magnetic barrier. Since the field of the terrestrial magnetic dipole falls off proportionally to the cube of the distance from the centre of the Earth, and since the inverted V structures are observed at a distance of about 1.5 terrestrial radii, the mirror ratio B_1/B_2 comes to 10–20; n is the electron concentration at the initial point and $v = (\frac{8}{\pi}T_e/m)^{1/2}$ is their thermal velocity. We see that a current sufficiently close to the maximum possible value

$$j_{max} = \frac{1}{4} env \frac{B_1}{B_2} \tag{9.10}$$

can pass through a sufficiently strong magnetic mirror $(B_1/B_2 = 10 - 20)$ (i.e., it can be greater than the random current by a factor of B_1/B_2, as a result of contraction of the magnetic flux), if the electrons traverse a potential difference which is sufficiently greater than the temperature. If no such potential difference is available, the maximum current density of the electron current from the far magnetosphere to the ionosphere cannot exceed the random current density $\frac{1}{4}env \simeq 10^{-6}\,\text{A/m}^2$ (typical values of parameters: $n = 0.1$–$1\,\text{cm}^{-3}$, $T_e = 0.3$–$1\,\text{keV}$). However, both observations [9.6] and numerical simulations [9.36, 9.37] indicate that the discrete auroras are generated only when the density of the electron current from the magnetosphere to the ionosphere becomes considerably greater than $10^{-6}\,\text{A/m}^2$ and satisfies relation (9.8) with a good safety margin. This is why it is of principal importance to identify the mechanism of generating an electric field which is capable of 'pulling' an electron current, which is greater than the random value (3.9), through a magnetic mirror.

In this section, I will discuss one of the hypotheses concerning the desired mechanism; it states that the mechanism is connected very closely with fluxes of positive ions which travel from the magnetospheric tail towards the ionosphere [9.6, 9.36–9.44]. The intensity of these fluxes and the ion energies grow considerably in 'violent' periods of the magnetosphere, for example, during magnetospheric substorms when (e.g., as a result of instability with respect to the reconnection of the magnetic lines of force [9.45–9.48]) strong induced electric fields are produced in the magnetosphere; these fields seem to help in the formation of the electrostatic fields that accelerate electrons which then excite polar auroras.

It is worthy of note that the average energy W_+ of the motion of ions along the magnetic field during such magnetospheric substorms reaches $10\,\text{keV}$ and greatly exceeds their temperature T_+ [9.6, 9.38–9.41]; note that the electron temperature $T_e \simeq 10^{-1}T_+$. This fact is connected with a theoretical model of formation of an electric double layer that has the potential difference of $e\Delta\varphi \simeq W_+ \gg T_e$, the positive polarity on the side of the ionosphere and the width

$$L \simeq (M/m)^{1/2}r_D \tag{9.11}$$

where r_D is the electron Debye radius. The model assumes (see [9.6, 9.40, 9.41, 9.43]) that the system includes five groups of charged particles: (1) fast

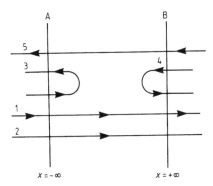

Figure 9.3   Five groups of charged particles in the theoretical model of the electric double layer in the auroral plasma [9.40]. Straight lines A and B trace arbitrary boundaries of the magnetospheric and the ionospheric sides of the double layer. 1, fast ions arriving from the magnetosphere and going through the double layer; 2, electrons arriving from the magnetosphere and freely transmitted through the layer; 3, 4, slow ions and electrons reflected back by the electric field of the layer; 5, ions accelerated by the field of the layer towards the magnetosphere.

magnetospheric ions which pass across this putative layer, (2) slow ions that enter the layer on the magnetospheric side and are reflected by the potential of the layer, (3) electrons which enter the layer on the magnetospheric side and are accelerated by the field of the layer towards the ionosphere, (4) slow electrons entering the layer on the ionospheric side and reflected back by its field, and, (5) slow ionospheric ions accelerated by the field of the layer towards the magnetosphere (see figure 9.3).

Numerical experiments have demonstrated that the interaction of these five groups of particles leads to the formation of the electric double layer with the above-mentioned parameters, located between the magnetosphere and the ionosphere; this layer is localized at a distance of one to one-and-a-half terrestrial radii from the Earth's surface. It is of principal importance to emphasize that the double layer appears in the system without applying any external potential difference to it: the value of $\Delta\varphi$ is mostly determined by the energy W_+ of the directed motion of magnetospheric ions. The conditions of stability of this double layer are definitely of a more complicated form than (9.2)–(9.7).

As a next step, the following geometric complication is introduced in comparison with figure 9.3: (a) it is assumed that the width of the beam of magnetospheric ions is limited to a value of about ten electronic Debye radii, less than the total width of the system, and (b) an external magnetic field, which magnetizes the electrons more strongly than it magnetizes the ions, is imposed on the system; as a result, a separation of charges in the transverse direction is produced. The solution of this problem gives the

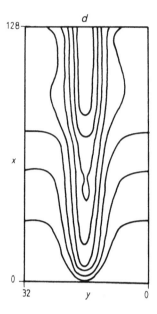

Figure 9.4   Potential distribution along the electric double layer (the direction x, from the ionosphere towards the magnetosphere) and across the layer (the direction y, to the left and to the right of the symmetry axis). The sizes along x and y are given in units of the Debye electron radius, r_D. The potential difference between the neighbouring curves is $\Delta\varphi = 0.07T_i$, where T_i is the temperature of ions arriving from the ionosphere [9.6, 9.41].

double layer structure shown in figure 9.4. Its effective width is 10 to 20 r_D, the effective length is of the order of $100r_D$. The double layer of the type shown in figure 9.4 can be referred to as an 'inverted V structure' of the type observed in the near-extraterrestrial plasma [9.5, 9.6].

Figure 9.5 obtained by numerical simulation [9.41] shows that if $T_+ \gtrsim T_e$, the potential difference in the layer is $\Delta\varphi \gtrsim T_+/e$, that is, of the order of several kV—just as in the observed inverted V structures.

The next step in the complexity of model [9.43] stems from the fact that magnetospheric ions moving towards the Earth meet in their path the 'magnetic mirror' of the terrestrial magnetic dipole. Since the distribution of ions is very broad, they are decelerated as they approach the Earth by the terrestrial magnetic barrier and a considerable fraction are reflected, thereby increasing the ion space charge density and creating the counter potential difference directed from the ionosphere to the magnetosphere. In principle, this phenomenon is related (although it is not identical) to the process of formation of the virtual anode in the ion beam. Electrons present in the system (both of magnetospheric and of ionospheric origin) partly compensate the ionic space charge, so that the extension of the region of

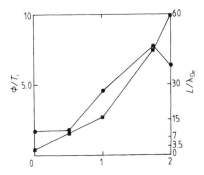

Figure 9.5   Potential difference and double layer thickness as a function of the ratio of the directed electron velocity to the thermal velocity at $T_e = 0.2T_i$. It is clear that $\varphi > T_i \gg T_e$ [9.41]. ● double layer potential; ■ double layer thickness.

the electrostatic field in question may greatly exceed the Debye radius of the incoming particles.

If we assume (as was done in [9.43]) that $W_+ \gg T_e$, then the potential difference in the evolving layer, $\Delta\varphi$, is considerably greater than T_e/e.

On the whole, the model described here looks quite realistic and pretends to offer a satisfactory qualitative interpretation of the inverted V structures observed.

It should be added here that observations [9.5, 9.26, 9.29, 9.50] point to the possibility of the formation of not one but a whole chain of weaker double layers resulting in the same integral effect of acceleration of particles.

The concept of formation of inverted V structures considered here is also supported by the result of an interesting experiment carried out by Stenzel *et al* [9.51–9.52]. This experiment imitates the situation of motion of an ion beam through the plasma in the model of the terrestrial magnetic dipole. Stenzel's experiment is schematically shown in figure 9.6.

The experiment was run in a large volume of plasma placed in a homogeneous magnetic field of several tens of Oe. A small permanent magnet producing field strength of several kOe was mounted inside this volume, so that its magnetic moment was opposite to the direction of the homogeneous magnetic field. A beam of Ar^+, Xe^+ and H_2^+ ions at energies of several tens of eV was injected into the system along the homogeneous magnetic field. Ions in general travelled towards the permanent magnet, bypassing it, but some of the ions went into one of its poles. It is of principal importance that ions were magnetized in the neighbourhood of the magnet: their Larmor radius was smaller than the transverse size of the beam. The geometry of this experiment thus resembled the magnetic configuration in the auroral zone of the Earth. The experiments produced a spectacular result: an electric double layer was formed in front of the magnetic pole

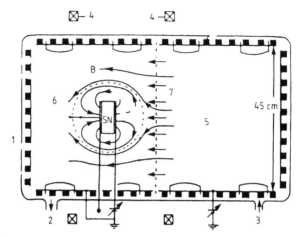

Figure 9.6 Schematic diagram of the experiment of Stenzel *et al* [9.51, 9.52]: 1, vacuum chamber filled with plasma; 2, to vacuum pump; 3, gas injection; 4, Helmholtz coils producing the homogeneous magnetic field; 5, ion beam plasma; 6, target plasma; 7, ion beam; 8, permanent magnet. Gas pressure $p = (2-20) \times 10^{-5}$ mm Hg. Plasma parameters: $n_e = 10^8$ cm^{-3}, $T_e = 2$ eV. The double layer is formed in the vicinity of the southern pole S of the magnet.

(sufficiently far away from its surface) where several ion Larmor radii can be placed on one beam diameter; the positive polarity of the double layer faces the magnet, its potential difference is close to the beam ion energy, and it has an inverted V shape, similar to that of the observed inverted V structures in the auroral zones of the ionosphere (i.e., in the zones of polar auroras). This experiment favours the standpoint, given above, of the 'ion-magnetospheric' mechanism of the formation of electric double layers above the ionosphere of the Earth.

In another experiment by the same group of authors [9.53], the phenomenon of reconnection of magnetic lines of force in the plasma was studied in the vicinity of a neutral plane, on both sides of which there are oppositely directed magnetic fields (a model of the equatorial region of the magnetospheric tail, see [9.33]). The reconnection of magnetic lines was triggered by passing a relatively low current through the plasma along the neutral plane. It was found that as a result of magnetic line reconnection, the induced electric field forms an electric double layer which accelerates ions along the neutral plane. Quite probably, this result is relevant to the problem of the source of particles that generate aurora borealis.

On the whole, the results of Stenzel's experiments are in good agreement with the predictions of the theory of formation of the inverted V structures outlined above (figure 9.2) and appear to give a better account of the finiteness of the ion beam and of the reflection of ions from the

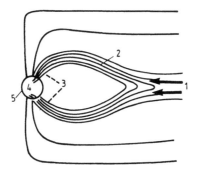

Figure 9.7   Schematic representation of initiation of discrete polar auroras by fast magnetospheric ions acting as their energy source. 1, fast ions moving from the magnetosphere tail; 2, magnetic force lines of terrestrial dipole; 3, regions of strong magnetic 'mirrors'; 4, the Earth; 5, Earth day side.

inhomogeneous field of the Earth's magnetic dipole.

The model picture of one of the probable mechanisms of discrete polar auroras outlined in this section can thus be illustrated by the schematic figure 9.7.

I will also mention the reviews [9.54] and [9.55] which discuss many of the phenomena treated above.

References

[9.1]   Block L P 1978 *Astrophys. and Space Phys* **55** 59

[9.2]   Kan J R and Lee L C 1980 *J. Geophys. Res.* **A85** 788

[9.3]   1982 *Symposium on Plasma Double Layers* Risö National Lab., Denmark

[9.4]   1984 *Second Symposium on Plasma Double Layers and Related Topics* (Insbruck, Austria)

[9.5]   Temerin M and Moser F S 1984 *Second Symposium on Plasma Double Layers and Related Topics* (Insbruck, Austria) p. 119

[9.6]   Akasofu S–I 1981 *Rept. Progr. Phys.* **44** 1123

[9.7]   1981 *Physics of Auroral Arc Formation* ed S-I Akasofu and J R Kan (Washington)

[9.8]   Sato T 1982 Auroral Physics *Magnetospheric Plasma Physics* ed A Nishida (Dordrecht: D Reidel) p. 197

[9.9]   Singh N and Schunk R W 1984 *Plasma Phys. Contr. Fusion* **26** 859

[9.10]   Singh N, Thiemann H and Schunk R W 1986 *IEEE Trans.* **PS-14** 805

[9.11]   Singh N, Thiemann H and Schunk R W 1987 *Planet Space Sci.* **35** 353

[9.12] Torven S, Lindberg L and Carpenter R T 1985 *Plasma Phys. Contr. Fusion* **27** 143

[9.13] Hairapetian G and Stenzel R L 1990 *Phys. Rev. Lett.* **65** 175

[9.14] Lindberg L 1988 *Astrophys. and Space Science* **144** 3

[9.15] Song B, Merlino R L, D'Angelo N 1990 *Phys. Fluids* **B2** 1936

[9.16] Cartier S L and Merlino R L 1987 *Phys. Fluids* **30** 2549

[9.17] Takeda Y and Yamagiva K 1985 *Phys. Rev. Lett.* **55** 711; 1991 *Phys. Fluids* **B3** 288

[9.18] Lutsenko E I, Sereda N D and Tseluiko A F 1988 *Zh. Tekhn. Fiz.* **58** 1299 (see also references therein) (Engl. Transl. 1988 *Sov. Phys. Tech. Phys.* **33** 1348)

[9.19] Iizuka S, Saeki K, Sato K and Hatta Y 1979 *Phys. Rev. Lett.* **13** 1404; 1985 *J. Phys. Soc. Japan* **54** 950

[9.20] Iizuka S, Michelsen P, Rasmussen J J, Schrittwieser R, Hatakeyama R, Saeki K and Sato N 1985 *J. Phys. Soc. Japan* **54** 2516

[9.21] Leung P, Wong A Y and Quon B H 1980 *Phys. Fluids* **25** 992

[9.22] Jovanovič D, Lynov J P, Michelsen P, Pécseli H L, Rasmussen J J and Thomsen K 1982 *Geophys. Res. Lett.* **9** 1049

[9.23] Cheung P K and Wong A Y 1987 *Phys. Rev. Lett.* **59** 551

[9.24] Cheung P K, Donovan S and Wong A Y 1988 *Phys. Rev. Lett.* **59** 1360

[9.25] Mozer F S, Catell C A, Hudson M K, Lysak R L, Temerin M and Torbert R B 1980 *Space Sci. Rev.* **27** 155

[9.26] Mozer F S, Boehm M H, Catell C A, Temerin M and Wygant J R 1985 *Space Sci. Rev.* **42** 313

[9.27] Torbert R B and Mozer F S 1978 *Geophys. Res. Lett.* **5** 135

[9.28] Bennett E L, Temerin M and Mozer F S 1983 *J. Geophys. Res.* **88A** 7107

[9.29] Redsun M S, Temerin M and Mozer F S 1985 *J. Geophys. Res.* **90A** 9615

[9.30] Kletzing C, Catell C, Mozer F S, Akasofu S–I and Makita K 1983 *J. Geophys. Res.* **A88** 4105

[9.31] Thieman J R and Hoffman R A 1985 *J. Geophys. Res.* **A90** 3511

[9.32] Wagner J S and Kan J R 1985 *Planet Space Sci.* **33** 89

[9.33] Lyons L R and Williams D J 1984 *Quantitative Aspects of Magnetospheric Physics* (Dordrecht: D Reidel)

[9.34] Antonova E E and Tverskoi B A 1975 *Geomagnetizm i Aeronomiya* **15** 105

[9.35] Knight L 1973 *Planet Space Sci.* **21** 741

[9.36] Yamamoto T and Kan J R 1985 *J. Geophys. Res.* **A90** 1553

[9.37] Yamamoto T and Kan J R 1986 *J. Geophys. Res.* **A91** 12113

[9.38] Swift D W 1978 *Space Sci. Rev.* **22** 35

[9.39] Kan J R 1982 *Space Sci. Rev.* **31** 71

[9.40] Williams A C 1986 *IEEE Trans.* **PS-14** 800

[9.41] Wagner J S, Tajima T, Kan J R, Leboeuf L N, Akasofu S–I and Dawson J M 1980 *Phys. Rev. Lett.* **45** 803

[9.42] Lyons L R and Evans D S 1984 *J. Geophys. Res.* **A89** 2395

[9.43] Serizawa Y and Sato T 1986 *Phys. Fluids* **29** 2753

[9.44] Frank A 1985 *Space Sci. Rev.* **42** 211

[9.45] Kadomtsev B B 1987 *Uspekhi Fiz. Nauk* **151** 3 (Engl. Transl. 1987 *Rep. Prog. Phys.* **50** 115)

[9.46] Kadomtsev B B 1979 *Nelineiniye Volny (Nonlinear Waves)* ed A V Gaponov–Grekhov (Moscow: Nauka) p. 131

[9.47] Syrovatsky S I 1978 *Priroda* **6** 84

[9.48] McPerron R 1979 *Rev. Geophys. Space Phys.* **17** 657

[9.49] Akasofu S–I and Chapman S 1972 *Solar-Terrestrial Physics* (Oxford:Clarendon)

[9.50] Temerin M, Gerny K, Lotko W and Mozer F S 1982 *Phys. Rev. Lett.* **48** 1175

[9.51] Nakamura Y and Stenzel R L 1982 *Symposium on Plasma Double Layers* Risö National Lab. (Denmark) p. 153

[9.52] Stenzel R L, Ooyama M and Nakamura Y 1981 *Phys. Fluids* **24** 708

[9.53] Stenzel R L, Gekelman and Wild N 1982 *Symposium on Plasma Double Layers* Risö National Lab. (Denmark) p. 181

[9.54] Raadu M A and Rasmussen J J 1988 *Astrophys. and Space Science* **144** 43

[9.55] Borovsky J E 1992 *Phys. Rev. Lett.* **69** 1054

10 Langmuir Solitons in Electron (Plasma) Beams

10.1 Solitons in Beam Physics

Only the linear theory was used in the preceding chapters of the book to analyse dispersion equations relevant for beam instabilities. This theory points to the conditions of generation of instabilities and predicts the frequencies and increments of oscillations which grow with time (in the neighbourhood of the instability threshold). However, the linear theory ignores those non-linear processes which limit the amplitude of waves and, generally speaking, predetermine their further evolution. The most effective among them are the following processes. (1) *Capture of beam particles by high-amplitude wave.* This phenomenon takes place when the potential oscillation amplitude φ_m of the wave times the particle charge e is greater than the kinetic energy of a beam particle in the frame of reference of the wave, $\frac{1}{2}m(u - v_{ph})^2$, where u is the velocity of a beam particle and $v_{ph} = \omega/k$ is the phase velocity of the wave. The particles captured by the wave oscillate in the wave's potential well, being successively reflected from its front and then its rear barriers (transferring energy to the wave and then draining it from the wave, respectively). Now the wave amplitude ceases to increase in time at the same rate and begins to go through (almost) periodic oscillations. However, the interaction of the beam with the waves it generates does not stop here. It was found [1.15, 1.16] that a new, so-called satellite instability now develops in the beam. Its characteristic frequency is smaller than the plasma frequency by the frequency of beam oscillations in the potential well of the fundamental wave. The amplitude of this satellite gradually increases and after some time it begins to act as the fundamental wave: it captures particles etc. The beam therefore continues to relax to a state with a plateau-like distribution function and then to a Maxwellian distribution. (In fact, treating problems related to this aspect goes far beyond the scope of this book.)

(2) *Wave scattering by plasma particles.* It is often convenient to treat

plasma waves (of frequency ω and wave vector k) in quantum mechanical terms, i.e., as an ensemble of plasmons whose energy is $\hbar\omega$ and momentum is $\hbar k$. Such plasmons can be scattered by plasma particles, reducing their momenta and thereby increasing the phase velocity ω/k of the wave. This process throws the wave out of synchronism with the beam particles ($v_{ph} > u$) and thus limits the wave amplitude. This process is an induced one, and its efficiency is proportional to the wave energy [10.1].

(3) There exists another non-linear process which, in contrast to the induced wave scattering by particles, does not reduce but increases the wave numbers of oscillations, that is, it reduces the phase velocities of the waves. This is known as the modulational wave instability [1.15, 1.16, 10.2–10.5]. The modulational instability produces a spatial redistribution of the energy of a quasi-monochromatic wave, in which a wave modulation, produced randomly by a fluctuation, begins to grow and the wave transforms into a wave packet. This wave packet then undergoes self-compression and the initially uniform spatial distribution of energy evolves to something like the deadly 'ninth wave' that seascape artists sometimes paint. Given favourable conditions, this instability produces single waves of high amplitude, known as *solitons*. The self-compression of wave packets increases the wave numbers of oscillations, that is, decreases their phase velocities, and also knocks the wave out of synchronism with beam particles ($v_{ph} < u$).

These processes may result in a remarkable transparency of the plasma, that is, in an increased beam relaxation length [1.16, 10.6, 10.7]. As an example, we can look at the following observation which seems to be caused by the modulational instability of Langmuir waves: the relaxation length of a flux of relativistic electrons of the solar wind in the near-solar plasma grows by several orders of magnitude and increases to more than a half of the distance from the Sun to the Earth [10.7, 10.8].

The theory of processes (1)–(3) achieved different degrees of success. Processes (1) and (2) have been analysed in sufficient detail [1.15, 1.16, 10.1] but the modulational instability (process (3)) and the soliton evolution are at the stage of intense investigation now. Consequently, the construction of a systematic theory cannot be regarded as brought to completion, and the theory of beam relaxation in the plasma is now being reborn.

Solitons arise not only as a result of modulational instability; they exist in various branches of plasma waves [10.3, 10.4, 10.9–10.19]. They play an essential role in the formation of the distribution function of the predominant part of electrons in the beam plasma; this is important, for example, for plasma chemistry [5.9, 10.20, 10.21] and for plasma heating techniques [10.22].

Solitons may produce a decisive effect on the evolution of the plasma turbulence [10.22–10.25] and on anomalous transfer processes in the plasma [10.17]; they may be directly relevant to the problem of new methods of charged particle acceleration [7.23]. Physically, solitons of some types are quite close to collisionless shock waves [1.16, 10.10]. At the present stage, the soliton is regarded as a fundamental object in non-linear wave theory. It is even said sometimes that the soliton in non-linear wave theory is an analogy to the classical oscillator in linear oscillation theory. Solitons play a principal role in modern theoretical physics [10.26–10.35] and even in theoretical bio- and neurophysics [10.26, 10.36].

Note that solitons are very often discovered 'at the tip of a pen' or as a result of computer simulations. It is therefore important to know how solitons behave (provided they do exist) in the real plasma with its inevitable spatial inhomogeneities, boundaries, extraneous waves, etc. The progress of the theory thus depends essentially on the experimental data. Since the main 'source' of solitons in the 'beam' plasma is the modulational instability of Langmuir waves, in this chapter I will focus attention on the experimental studies of this instability and of solitons that it produces. A brief theoretical introduction at a qualitative level will preceed the presentation of these experimental data.

As for the history of the soliton, it is very briefly this. The 160th anniversary of the discovery of solitons on the so-called 'shallow water', whose depth d is less than the wavelength λ (Scott–Russell), comes in 1994, and in 1995 we will have the centenary of the first sufficiently general theoretical description by Korteweg and de Vries (KDV) in whose work a solitary wave (the soliton) was presented for the first time as a result of equilibrium of dispersion and non-linearity (see [10.26], p 215). In plasma physics, solitons were first described in 1958 in a theoretical paper by Sagdeev [10.37] and were later studied in [10.9, 10.10] where ion-acoustic and magneto-acoustic solitons (of the KDV type) were theoretically treated. Experimentally, solitons of this type were first observed in a plasma by Ikezi [10.38] who also carried out very interesting experiments on soliton collisions. The term 'soliton' (from 'solitary wave') became widespread largely owing to the paper of Zabusky and Kruskal (see [10.26], p. 217).

Solitons of another type, the so-called 'deep water solitons' ($d > \lambda$), arise as a result of the modulational instability of waves. This instability was first predicted in 1965 by Lighthill [10.39] and discovered by Benjamin and Feir in experiments on water [10.40, 10.41]. The modulational instability of Langmuir waves was predicted in 1964 by Vedenov and Rudakov [10.2]. A similar instability of a laser beam in a non-linear dispersive medium results in its three-dimensional focusing and produces a sequence of moving foci (solitons) (see the publications of Askaryan, Talanov, Townes and their co-workers [10.42–10.44] and the review [10.45]). The problem of soliton stability was analysed theoretically by Kadomtsev and Petviashvili [1.15]. A recent review of this field can be found in [10.46].

A pioneering theoretical work on Langmuir wave collapse was presented by Zakharov [10.47], and on Langmuir solitons, by Rudakov [10.48].

It is found that such natural phenomena as the largest and longest-lived vortices in the oceans and in planetary atmospheres (e.g., Jupiter's famous vortex 'Great Red Spot', which has already been observed for more than three centuries) can be treated as Rossby solitons; these are objects that are physically analogous to drift solitons in magnetized plasma (see, e.g. [10.49, 10.50]).

10.2 Physical Concept of Self-Compression (Collapse) of Langmuir Waves and Soliton Formation

High-amplitude plasma waves (non-linear waves) manifest, under certain conditions, the modulational instability which results in their self-compression. Before discussing this instability, it will be useful to discuss a number of notions which are to be used to varying degrees in the discussion to follow.

One of these notions is wave decay, including the decay of a Langmuir wave to a Langmuir and an ion-acoustic wave [1.15, 1.16, 10.1]. The wave vector of the original wave is k and the frequency, corresponding to the dispersion of Langmuir waves, is

$$\omega^2 = \omega_p^2 + 3k^2 v_e^2 = \omega_p^2 + 3k^2 T_e/m \tag{10.1}$$

that is, $\omega_k = \omega_p + \frac{3}{2}k^2 r_D^2 \omega_p$; those of the secondary Langmuir waves are k' and $\omega_k' = \omega_p + \frac{3}{2}(k')^2 r_D^2 \omega_p$, respectively, and those of the ion-sound wave are k_s and $\omega = k_s c_s$, where c_s is the velocity of ion sound. These waves are an ensemble of quanta with energy $\hbar\omega_i$ and momentum $\hbar k_i$ $(i = k, k', s)$; the laws of energy and momentum conservation for this particular decay are

$$\hbar\omega_k = \hbar\omega_k' + \hbar\omega_s$$
$$k = k' + k_s.$$

Let $k' \| k_s \| k$. Taking into account the functions $\omega_i(k_i)$, these equations immediately yield

$$2k > (k + k') = \frac{2}{3}\frac{(m/M)^{1/2}}{r_D}$$

or

$$k r_D > \tfrac{1}{3}(m/M)^{1/2}. \tag{10.2}$$

This is the condition allowing the decay. Correspondingly, the condition forbidding the decay is

$$k r_D < \tfrac{1}{3}(m/M)^{1/2}. \tag{10.3}$$

An important channel of decay is that of backscattering of the Langmuir wave, in which $k' = -k$, $k_s = 2k$. For this case the conservation laws obviously give

$$(k - k') r_D = \tfrac{2}{3}(m/M)^{1/2}$$

whence we find that the minimal wave number of the original wave must be greater than in the first example of decay.

Another important notion is the force of dynamic pressure, which pushes out the plasma from the region of higher strength of the electric field oscillations [10.51] (see also [1.15, 1.16, 10.3]). This force is sometimes called the Miller force [1.15]. I will use the expression for this force, which is valid in the so-called quasistatic approximation, in which the group velocity of the (Langmuir) waves is much lower than the ion-sound velocity:

$$\frac{\partial \omega}{\partial k} = 3kr_D\omega_p r_D \ll c_s = \omega_p r_D (m/M)^{1/2} \tag{10.4}$$

or

$$kr_D \ll \tfrac{1}{3}(m/M)^{1/2}. \tag{10.5}$$

This approximation is qualitatively equivalent to condition (10.3) of the impossibility of the Langmuir wave decay. In this approximation, the volumetric density of the dynamic pressure force is

$$F = -n\nabla \frac{e^2 E_0^2}{4m\omega^2}$$

that is, for the Langmuir waves it gives

$$F = -\nabla W_e = -n\nabla T_\sim \tag{10.6}$$

where $W_e = E_0^2/16\pi$ is the electric energy density of the waves whose amplitude is E_0, and $nT_\sim = nmv_\sim^2/2$ is the mechanical energy density of the electrons which oscillate in the wave. It was assumed in (10.6) that the electric and mechanical energy densities in the Langmuir wave are equal (see section 4.1):

$$nT_\sim = nmv_\sim^2/2 = W_e = E_0^2/16\pi$$

where $v_\sim = eE_0/m\omega_p$ is the amplitude of the velocity of the oscillating motion of an electron in the wave. The force F causes density 'wells' in the plasma; the depth δn of a well is determined, for condition (10.5) satisfied and $T_i \ll T_e$, by the equilibrium between the force F and the excess of thermal pressure of the plasma, caused by the formation of the well:

$$F = -\nabla W_e = -n\nabla T_\sim = \nabla(nT)$$

that is,

$$\frac{\delta n}{n} = \frac{\Delta W_e}{nT} = \frac{\Delta T_\sim}{T} \tag{10.7}$$

where $T = T_e$ and ΔW_e is the density excess of the electric energy (or of the mechanical energy, which is an equivalent statement) of the wave in the density well, as compared with its value in the surrounding plasma. In what follows, we briefly recast relation (10.7) to the form

$$\frac{\delta n}{n} = \frac{W}{nT} = \frac{T_\sim}{T} \tag{10.8}$$

where W and $T_\sim$ are understood as the magnitudes of ΔW_e and $\Delta T_\sim$, respectively.

The dynamic pressure force thus 'digs' a density well in the plasma. It is not difficult to see that this well will concentrate the Langmuir wave, that is, will suck in the plasmons (the quanta of the Langmuir field) [1.15, 10.52]. Any wave is characterized by its phase $\varphi = \omega t - kr$, the frequency $\omega = \partial\varphi/\partial t$ and the wave vector $k = -\nabla\varphi$, so that

$$\frac{\partial k}{\partial t} = -\frac{\partial}{\partial t}\nabla\varphi = -\nabla\frac{\partial\varphi}{\partial t} = -\nabla\omega$$

or $\partial\hbar k/\partial t = -\nabla(\hbar\omega)$; therefore, at the Langmuir frequency we have

$$\frac{\partial\hbar k}{\partial t} = -\hbar\nabla\omega_p. \tag{10.9}$$

Since $\hbar k$ is the plasmon momentum, the last equation is the equation of motion of a plasmon subjected to the force $-\hbar\nabla\omega_p$, which is caused by the density gradient in the well and is directed towards the well†. The volumetric density of this force can be expressed by defining the Langmuir plasmon density

$$N = \frac{W_e + nT_\sim}{\hbar\omega_p}$$

which is the ratio of the wave energy density (i.e., the sum of the electric and mechanical energies) to the energy of one plasmon. Using the obvious relation

$$\frac{\delta\omega_p}{\omega_p} = -\frac{1}{2}\frac{\delta n}{n} = -\frac{1}{2}\frac{W}{nT} \tag{10.10}$$

we arrive at the volumetric density of the force which sucks plasmons into the density well:

$$N\frac{\partial}{\partial t}(\hbar k) = -N\hbar\nabla\omega_p = -T_\sim\nabla n = -W_e\frac{\nabla n}{n}.$$

Therefore, if a density well is randomly produced in the plasma, the plasma partly captures the wave, increasing its energy density in the well. The force

† This is the physical cause of the 'nucleation' of Langmuir waves on the plasma density wells, described in detail in [10.24].

(10.6) that increases the depth of the well is thereby increased, which results in further concentration of the wave energy in the well, and so forth. This is the physical meaning of the modulational instability. In view of these arguments, the frequency of the non-linear Langmuir wave is

$$\omega = \omega_p + \frac{3}{2}(kr_D)^2\omega_p - \frac{W}{2nT}\omega_p. \tag{10.11}$$

It is now possible to give an interpretation, without claiming impeccable rigour, of the physical meaning of the modulational instability in terms of decay processes. A monochromatic wave under modulational instability becomes amplitude-modulated and its spectrum gets more complicated: before modulation, it consisted of a single frequency ω_k but after modulation it consists of the main frequency $\omega_{k'}$ and two satellite lines symmetrically shifted with respect to $\omega_{k'}$. The wave numbers of the satellites differ from the wave number of the main frequency by the modulation wave number $\kappa = 2\pi/L$, where L is the modulation wavelength, and the satellite amplitudes at the modulation coefficient $m = 1$ equal one half of the amplitude of the main wave. As for the main wave frequency $\omega_{k'}$ after modulation, we can now assume, on the basis of the above analysis, that it deviates from the original wave frequency ω_k by the non-linear frequency shift (10.10). Therefore, the energy conservation law for the Langmuir wave modulation can be written as

$$\left.\begin{array}{c} \hbar\omega_k + \hbar\omega_k = \hbar\omega_k' + \frac{1}{2}\hbar\omega_{k-\kappa} + \frac{1}{2}\hbar\omega_{k+\kappa} \\[2mm] 2\delta\omega = W\omega_p/nT = \frac{3}{2}(\kappa r_D)^2\omega_p \end{array}\right\}. \tag{10.12}$$

or

Therefore, for the modulation to occur, the Langmuir wave energy density must exceed the threshold value $(W/nT)_{th}$:

$$\frac{W}{nT} \gtrsim \left(\frac{W}{nT}\right)_{th} = \frac{3}{2}(\kappa r_D)^2. \tag{10.13}$$

The result obtained here is independent of the wave modulation coefficient: if it is below unity, the satellite amplitudes are lower but the process remains qualitatively unchanged: only the ratio of the number of basic quanta and that of satellite quanta is altered. Relation (10.13) determines the modulation period $L = 2\pi/\kappa$ as a function of wave energy density: the higher W, the lower L. Note (almost jokingly) that if $L = 9\lambda$, where $\lambda = 2\pi/k$ is the main wavelength, the modulational instability will produce the 'ninth wave'-type picture; this will take place at a wave energy density W_9, such that

$$\frac{W_9}{nT} = \frac{1}{54}(kr_D)^2. \tag{10.14}$$

If $W < W_9$, the large waves follow one another less frequently, and if $W > W_9$, they are more frequent. Process (10.12) is known as second-order decay [10.53]. It is very special in that it is realized, for example, when condition (10.3) is satisfied, so that the ordinary decay discussed above is forbidden.

It is important to emphasize, however, that the energy criterion (10.13) is necessary but not sufficient condition for the modulational instability. A second necessary condition is the so-called Lighthill criterion [1.15, 10.3, 10.42, 10.54]

$$\alpha \frac{\partial v_g}{\partial k} \equiv \alpha \frac{\partial^2 \omega}{\partial k^2} < 0 \qquad (10.15)$$

where α is the coefficient characterizing the non-linear wave frequency shift:

$$\omega = \omega_p + \tfrac{3}{2} k^2 r_D^2 \omega_p + \alpha E_0^2. \qquad (10.16)$$

In accordance with (10.16) and (10.11), we obtain

$$\alpha = -\frac{\omega_p}{32\pi n T}. \qquad (10.17)$$

The Lighthill criterion (10.15) follows from the so-called 'non-linear Schrödinger equation', which is a non-linear generalization of the parabolic Leontovich equation of diffraction theory [1.15], used to analyse the stability of wave packets. This criterion is valid only in the approximation in which the concept of the wave packet is meaningful, namely, if the size of the wave bunch is much longer than the wavelength of the original wave, that is,

$$L \gg \lambda \qquad \text{or} \qquad \kappa \ll k. \qquad (10.18)$$

For Langmuir waves, the Lighthill criterion is satisfied if the group velocity of the waves is not greater than the ion-sound velocity, that is, if the 'anti-decay' condition (10.3) is satisfied. In this case, as has been mentioned earlier, the non-linear shift of the Langmuir wave frequency is negative (formula (10.11)); furthermore, since $\partial^2 \omega / \partial k^2 > 0$, the product $\alpha \, \partial^2 \omega / \partial k^2$ is negative (in accordance with (10.15)).

The non-linear parabolic equation also yields the energy threshold of modulational instability which coincides with (10.13):

$$4|\alpha|E_0^2 \geqslant \left| \frac{\partial v_g}{\partial k} \right| \kappa^2 \qquad \text{or} \qquad \frac{W}{nT} \geqslant \frac{3}{2} (\kappa r_D)^2. \qquad (10.19)$$

The instability considered above is the property of a non-linear wave packet, which is a principal distinction from a linear packet. Indeed, it is well known that a linear wave packet (i.e. low-amplitude packet) spreads out rapidly. The rate of spreading is determined by the group velocity dispersion

$$\frac{\partial L}{\partial t} \simeq \left|\frac{\partial v_g}{\partial k}\right| 2\kappa \tag{10.20}$$

where L is the packet length and 2κ is the range of its constituent wave numbers. The characteristic time of packet spreading is

$$\tau = \frac{L}{\partial L/\partial t} \simeq \frac{\pi}{(\partial v_g/\partial k)\kappa^2}. \tag{10.21}$$

We saw, however, that if the packet is non-linear, it does not spread out; on the contrary, it undergoes self-modulation, that is, compression. Condition (10.19), which allows this, has a lucid physical meaning: the self-compression increment (which is approximately equal to the non-linear frequency shift)

$$\gamma \simeq \omega_p \frac{W}{nT} \tag{10.22}$$

is greater than the inverse spreading time $1/\tau$.

The modulational instability may generate a Langmuir soliton. Let us see how this mechanism works.

Let a packet of Langmuir waves of size $L \simeq 2\pi/\kappa$ contain a sufficiently large field energy which 'more than satisfies' the modulational instability condition (10.19). The packet then starts to collapse while its energy is conserved (provided no external pumping is at work): $WL = $ constant, that is, the energy density increases with decreasing L as $W \propto L^{-1}$. Since the additional dispersion term in the oscillation frequency, $\frac{3}{2}\kappa^2 r_D^2 \omega_p$, then increases proportionally to L^{-2} (i.e., steeper than W), equality will set in at a certain L^* instead of the initial inequality (10.19) and the collapse will be terminated. In other words, the non-linearity (packet collapse) dominates at $L > L^*$, while the dispersion (packet spreading) dominates at $L < L^*$; the non-linearity and dispersion effects balance out at $L = L^*$. In the equilibrium state, a self-compressed wave packet is formed whose envelope shapes the soliton: this is the so-called *envelope soliton*—a solitary wave with 'Langmuir filling', or a plasma density well which is 'filled' with Langmuir oscillations.

It is easy to see that such an equilibrium is possible only in the one-dimensional case. Indeed, in the case of two dimensions, we have $W_\sim \propto L^{-2}$, and in 3D, we have $W \propto L^{-3}$; hence, since the dispersion correction to frequency is again proportional to L^{-2}, non-linearity is not balanced out by dispersion and the packet collapses.

The electric field of a Langmuir soliton (see [10.22]) is

$$E(z,t) = \frac{E_0}{\cosh[k_0(z - ut)]} \exp\{i(kz - \omega t)\} \tag{10.23}$$

where z is the longitudinal coordinate.

The oscillation frequency of the soliton is

$$\omega = \omega_p + \frac{3}{2}r_D^2(k^2 - k_0^2)\omega_p \qquad (10.24)$$

the wave number is

$$k_0 = r_D^{-1}\left(\frac{W}{6nT}\right)^{1/2} \qquad (10.25)$$

the propagation velocity is

$$u = \frac{\partial\omega}{\partial k} = 3\omega_p r_D^2 k < c_s \qquad (10.26)$$

and the density well depth is

$$\frac{\delta n}{n} \simeq \frac{W}{nT}.$$

The width of the soliton at a height of $1/e$ of the field amplitude $E = E_0 \cosh^{-1}(k_0 z)$ is†

$$\Delta_E \simeq \frac{3.3}{k_0} \simeq r_D\left(\frac{60nT}{W}\right)^{1/2} = r_D\left(\frac{60n}{\delta n}\right)^{1/2}. \qquad (10.27)$$

The width of the soliton at the height of $1/e$ of the field intensity maximum $E^2 = E_0^2 \cosh^{-2}(k_0 z)$ is

$$\Delta_{E^2} \simeq \frac{2}{3}\Delta_E \simeq \frac{2.2}{k_0} \simeq r_D\left(\frac{30nT}{W}\right)^{1/2} = r_D\left(\frac{30n}{\delta n}\right)^{1/2}. \qquad (10.28)$$

The width of the plasma density well is

$$\Delta_n = \Delta_{E^2} \simeq \frac{2}{3}\Delta_E \simeq r_D\left(\frac{30n}{\delta n}\right)^{1/2}. \qquad (10.29)$$

The size Δ characterizes the scale over which the non-linear compression of the wave packet is balanced out by its dispersion spreading.

The envelope soliton considered, in which

$$\Delta > \lambda \text{ and } \frac{W}{nT} \simeq \frac{\delta n}{n} \gtrsim (\kappa r_D)^2 < (k r_D)^2 < \frac{m}{M} \qquad (10.30)$$

is a non-spreading Langmuir wave packet. In contrast to a linear wave packet, the amplitudes and phases of wave components in the envelope soliton are correlated, so that the packet does not spread out.

† Formulas (10.27)–(10.29) were derived in [10.55]).

As follows from relations (10.24) and (10.25), the frequency reduction $\Delta\omega/\omega_p$ at $k \ll k_0$ is half that in the non-solitary Langmuir wave, namely

$$\frac{\Delta\omega}{\omega_p} = \frac{1}{4}\frac{\delta n}{n}.$$

The reason for this is that the frequency reduction $\frac{1}{2}\delta n/n$ is half-compensated for by the frequency increase due to the spatial localization of the wave.

The envelope soliton is a fascinating object of principal importance. However, owing to its relatively long extension ($\Delta \gg r_D$) and low oscillation intensity (10.30), this soliton is definitely not the best example of the situation which corresponds to the concept of the physical collapse of waves. Of much greater interest from this point of view is the large-amplitude Langmuir soliton formed in the regime of

$$\frac{W}{nT} > (kr_D)^2 \qquad \frac{W}{nT} > \frac{m}{M} \qquad (10.31)$$

whose size is less than the initial oscillation wavelength:

$$\Delta < \lambda = \frac{2\pi}{k}. \qquad (10.32)$$

This is the soliton which is usually referred to as the Langmuir soliton. This soliton is also characterized by properties (10.23)–(10.29). However, as a result of the formation of a very deep density well, the non-linear decrease in the oscillation frequency is now greater than the dispersion frequency correction (the absolute value of the third term in the right-hand part of (10.11) is greater than that of the second term) and $\omega < \omega_p$, that is, the oscillation frequency in the soliton is lower than the Langmuir frequency of the surrounding plasma. As a result, the Langmuir oscillation is trapped in the density well as if in a resonator cavity. (In the envelope soliton, $\omega > \omega_p$ and no trapping of the oscillation field occurs.) Since the size of a large-amplitude soliton is less than the initial wavelength ($k_0 > k$), the 'non-linear' dispersion is dictated by k_0 and, for obvious reasons, the soliton propagation velocity is not connected, by virtue of (10.26), to the wave number of the initial wave from which the soliton evolved. In the new situation, the soliton velocity u is a free parameter subject only to the condition $u < c_s$, and relation (10.26) can be regarded as a definition of that value of k that has to be substituted, for a given value of u, into (10.24) to obtain the exact oscillation frequency value.

The Langmuir soliton arising in the regime (10.31) is formed owing to the short-wavelength modulational instability of Langmuir waves with respect to self-compression. This instability has essentially the same physical meaning as the above long-wavelength modulational instability which generates the envelope soliton. Let us look more closely at the conditions of its formation. Let a Langmuir wave at a frequency $\omega_k = \omega_p + \frac{3}{2}(kr_D)^2\omega_p$ be in an external HF pumping field at a frequency ω_0, energy density W and spatial period $L \gg 1/k$. Then the wave amplitude will grow in time (aperiodically, with the increment of the order of the ion-sound frequency), provided [10.56–10.58]

$$\left.\begin{array}{c} \omega_0 \lesssim \omega_k \\[2mm] \dfrac{W}{nT} > \dfrac{\omega_k - \omega_0}{\omega_p} \end{array}\right\} . \tag{10.33}$$

Let us now apply these relations to determining the threshold of the aperiodic instability of the free Langmuir wave (in zero external field) with respect to the (low-frequency) modulation whose wave number $\kappa \gtrsim k$. Since this modulation increases the dispersion correction to the Langmuir wave frequency by $\frac{3}{2}(\kappa r_D)^2\omega_p$), formula (10.33) implies that the instability threshold increases:

$$\frac{W}{nT} \simeq \frac{3}{2}(\kappa r_D)^2 \gtrsim (kr_D)^2. \tag{10.34}$$

The strength of this inequality is determined by the relation between κ and k, that is, by the ultimate size of the soliton and the initial wavelength. Therefore, the long-wavelength modulational instability develops at $\kappa < k$ and has a relatively low threshold (10.19), while the short-wavelength variant of the modulational instability takes place at $\kappa > k$ and is characterized by a relatively high threshold (10.34). In this instability, the arising wave with wave number κ is found to be modulated by the longer-wavelength original wave ($k < \kappa$) and subsequently undergoes the self-compression which ultimately produces Langmuir solitons (in one-dimensional geometry).

If the original wavelength is very large (e.g., $\lambda \to \infty$), the wavefield splits into separate regions with a characteristic scale λ_M, for which the modulational instability increment is maximal; inside these fields, the wave will 'collapse' into solitons. This scale is determined by the relation [10.47]

$$\lambda_M \simeq 2\pi r_D \left(\frac{nT}{W_0}\frac{M}{m}\right)^{1/4}. \tag{10.35}$$

This situation is realized, for example, at the plasma resonance point when the external pumping wave is transformed to the Langmuir wave [5.39].

If, however, the original wave has a very limited wavelength λ, the characteristic scale λ_M may be imposed by λ. This regime of modulational instability develops when Langmuir waves are pumped by the electron beam [10.59–10.62] and

$$\frac{\lambda}{2} \leqslant \lambda_M \leqslant \lambda$$

where $\lambda = u/f_p$ is the Čerenkov wavelength. Soliton amplitude E_0 is determined by the conditions of pumping and is limited by Landau damping at such small sizes δ (large k_0), at which soliton width is of the order of electron Debye radius (see [10.25]).

The difference between the two types of modulational instability we are now discussing is not very meaningful if $k \to 0$, and also if we treat not a quasimonochromatic wave but a 'gas' of Langmuir plasmons whose wave numbers are spread within $\Delta k \simeq k$, since the long-wavelength modulational instability also has a threshold in this case (see (10.34)). It is of principal importance to bear in mind that the Lighthill criterion (10.15) holds only for the long-wavelength modulational instability of a wave packet whose size L is much greater than the original wavelength λ. As for the short-wavelength modulational instability in which a soliton of size $L \lesssim \lambda$ is formed and there is no Langmuir filling, it does not fall under criterion (10.15).

It is of interest to compare the behaviour of the Langmuir soliton with that of a linear packet of Langmuir waves of the same size and propagation velocity ($\lesssim c_S$). Assume that according to the experimental data (see section 10.4), this size is $\Delta \simeq 3/k_0 \simeq 20r_D \simeq 1\,\text{cm}$. In hydrogen plasma at density $n \simeq 3 \times 10^9\,\text{cm}^{-3}$ and temperature $T_e \simeq 10\,\text{eV}$, relation (10.21) predicts that a linear packet has a 'free path' $l \lesssim \tau c_S$ (until spreading to twice the original width) not greater than the initial size Δ. A soliton of a size of the order of $20r_D$ (which corresponds to the well depth $\delta n/n \simeq 5\%$) should, however, propagate without spreading.

Under certain conditions, a one-dimensional density well formed by intensive external HF field in the plasma contains not one but a large number of solitons [5.39]. This object, whose size is much greater than the width (10.27) of an individual soliton, is known as a *caviton*. According to the theory, a one-dimensional soliton in zero magnetic field must undergo rapid collapse in transverse directions [10.62]. Short-lived solitons are known as *spikons* [10.22].

Let us consider now the problem of experimental observation of the excitation of Langmuir solitons by electron beam.

The possibility of observing the Langmuir envelope soliton is greatly limited by its low amplitude. Indeed, equation (10.30) implies that the depth of the density well of the envelope soliton $\delta n/n < m/M$; this well cannot be detected against the background of the inevitable plasma noise of different origin. Consequently, only the high-amplitude soliton, formed by the short-wave modulational instability, is of interest for a comparison with experimental observations. Let us now find a criterion for its excitation by the electron beam. In order to make use of condition (10.34), we need to connect the maximum possible energy density of Langmuir waves $W = E_0^2/16\pi$ in the beam–plasma system, and the quantity $(kr_D)^2$ with the system's parameters. The relation $k = \omega_p/u$ implies $(kr_D)^2 = T_e/mu^2 =$

$T_e/2W_1$, so that the instability threshold is

$$\left(\frac{W}{nT}\right)_c = \frac{T_e}{2W_1}. \tag{10.36}$$

In order to find the maximum possible amplitude E_0 of the electric field of the wave, we will choose the following line of argument. Beam electrons are moving at a velocity which is approximately equal to the phase velocity of the wave. They are therefore captured by a wave of a sufficiently high amplitude and oscillate in it. The frequency of these oscillations will be found from the equation of motion of the captured electron (in the wave's frame of reference):

$$m\ddot{z} = -eE_0 \sin kz$$

where z is the longitudinal coordinate. In the linear approximation $\sin kz \simeq kz$, that is, $\ddot{z} + ekE_0z/m = 0$, and the oscillation frequency (known as 'bounce' frequency) is

$$\omega_b = \left(\frac{ekE_0}{m}\right)^{1/2} = \left(\frac{eE_0}{m}\frac{\omega_p}{u}\right)^{1/2}.$$

As long as beam particles are not captured by the wave, the monoenergetic beam is pumping the wave at an increment (3.34):

$$\gamma = \omega_p(n_1/n)^{1/3}$$

where n_1 is the beam density (we assume that $k_z = k$). As a result of the beam capture by the wave, wave pumping almost stops: this occurs at a wave amplitude at which the frequency of oscillations of captured electrons ω_b becomes almost equal to the beam instability increment γ. Indeed, if $\omega_b > \gamma$, the captured particles would be reflected many times from the front and the rear potential barriers of the wave over a time of about $1/\gamma$, and the averaged energy exchange between particles and the wave (over one period of particle oscillations in the potential well) would be almost zero. In this case, both the beam generation of the Langmuir wave and the Landau damping of the monochromatic wave in the 'ordinary' plasma fail to occur [1.15]. Therefore, the maximum electric energy density of the Langmuir wave is found by equating the frequency ω_b to the increment γ:

$$\frac{E_0^2}{16\pi} = \frac{1}{2}n_1\frac{mu^2}{2}\left(\frac{n_1}{n}\right)^{1/3}.$$

A combination of this result with the modulational instability condition $W/nT \gtrsim (kr_D)^2$ gives an expression for its threshold:

$$\frac{mu^2/2}{T_e} \gtrsim \left(\frac{n}{n_1}\right)^{2/3}$$

or

$$\frac{n_1 mu^2/2}{nT} \geqslant \left(\frac{n_1}{n}\right)^{1/3}. \tag{10.37}$$

For example, if $n_1 \simeq 10^{-2}n$ (which is close to the conditions of the experiments to be discussed below), then the beam electron energy must be bounded from below:

$$W_1 \equiv \frac{mu^2}{2} \geqslant 25T_e \tag{10.38}$$

so that $W_1 \geqslant 250\,\text{eV}$ at $T_e \simeq 10\,\text{eV}$, and $W_1 \geqslant 25\,\text{eV}$ at $T_e = 1\,\text{eV}$. These conditions were readily met in the experiments outlined below. The short-wavelength modulational instability is often referred to in the literature as the *oscillating two-stream instability* (OTSI) [10.64].

Relation (10.37) is easily transformed to

$$\frac{n_1}{n} \geqslant \left(\frac{T_e}{mu_0^2/2}\right)^{3/2}. \tag{10.37'}$$

The equality sign yields the threshold of the modulational instability generated by a non-monoenergetic electron beam. If we assume that the beam has considerable velocity spread, it is not difficult to achieve the exponent 2 instead of 3/2 in the expression in the right-hand side of (10.37'). It is then possible to speak of a qualitative agreement of the theory and the experiment of [10.63], which carefully analysed the threshold condition of modulational instability of Langmuir waves, generated by a non-monoenergetic electron beam in a non-magnetized plasma.

From the standpoint of the theory, solitons may be either moving or stationary. It was shown (see [10.65]) that a high-amplitude Langmuir soliton (of a size of only several Debye radii) converts into a standing soliton when plasma quasineutrality in it is violated.

The difference between low- and high-amplitude solitons is found to be of an essentially qualitative, not simply quantitative nature. According to computer calculations [10.22, 10.66], envelope solitons are not affected by collisions, while high-amplitude solitons tend to merge. In this last case, the energy excess implied by conservation laws is liberated in the form of ion acoustic waves which, in their turn, produce soliton fragmentation. All these processes lead to strong Langmuir turbulence.

It is of interest to recall that when the dynamics of large-amplitude Langmuir solitons was not yet known, some authors (e.g., see the first review paper on solitons [10.26]) tended to include in the definition of the soliton the property of being unchanged after collisions; we now know that not all solitons possess it. In view of what I have said above, this definition is a matter of 'taste'.

10.3 Langmuir Cavitons in Non-Magnetized Plasma (Experiment)

The formation of non-linear waves of the type of Langmuir cavitons (plasma-density wells filled with Langmuir waves) was first observed in the experiments of [5.39] where Langmuir waves were generated by the electric field of an electromagnetic wave which was incident on non-uniform plasma in zero magnetic field. The plasma density n was 'sufficiently high': the Langmuir frequency corresponding to the density maximum, $\omega_p(n_{max})$, was higher than the wave frequency ω. The plasma was thus opaque to the wave, and the wave was reflected from the plasma in the layer where $\omega = \omega_p(n)$. Since the wave was incident on the plasma not quite perpendicularly to the plasma surface, it had a non-zero normal component of the electric field. In this case, the wave energy is partly transformed into that of Langmuir waves. The Langmuir waves produced density wells in the plasma and were localized in them. The electric field of Langmuir waves was measured by 'probing' the plasma with a transverse electron beam (see section 5.6); it was shown that the depth of the well was proportional to the Langmuir wave energy density:

$$\frac{\delta n}{n} \simeq \frac{W}{nT} = \frac{E_0^2}{16\pi nT}.$$

The value of $\delta n/n$ reached 0.2 and even higher, with the width of the density well being $(50\text{--}100)\, r_D$. The distributions of the electric field and of the plasma density in the caviton are shown in figure 10.1. As follows from the theory presented in the preceding section (see relation (10.27), the well width Δn at $\delta n/n = 0.2$ and with only one soliton localized in the well should be about $12r_D$, that is, it must be much less than the observed value. This means that each caviton in the experiments of [5.39] contained a large number of Langmuir solitons which overlap and cannot be seen separately. An experiment which was very similar to [5.39] both in design and in its results was reported in [10.67]. Qualitatively analogous results were reported in some later papers, for instance [10.68–10.72].

The discussion below will be limited to those experiments in which electron beams were used to generate Langmuir waves in the plasma and to create non-linear soliton-type structures. These experiments can be divided into two groups: in the first group ([10.59, 10.69, 10.73]) no magnetic field was imposed ($H = 0$), while in the second group ([10.60, 10.61, 10.74, 10.75]) the Langmuir oscillations were magnetized: the Larmor frequency was higher than the Langmuir frequency ω_p. Let us look at these experiments in the order given above. The self-compression of the Langmuir waves in non-magnetized beam plasma was first observed in [10.59] (see also [10.76]). The experiments were set up as follows. A plasma column of about 30 cm in diameter and about 100 cm in length was produced in a large evacuated volume about 2 m across, using a large number of electron

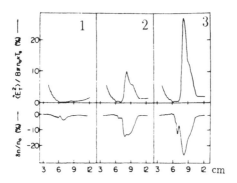

Figure 10.1 Spatial distributions of electric field energy and plasma density in the region of the plasma resonance, at three times after the pump wave was switched on. Horizontal coordinate: distance from the pump wave source [5.39]. 1, $t = 1\,\mu s$; 2, $t = 6\,\mu s$; 3, $t = 10\,\mu s$.

sources (see figure .9.7). Grids divided the plasma column into two parts (hence the name for the set-up: the 'double plasma device'), with a potential difference between them. This potential difference drove a flow of electrons predominantly from the 'negative' to the 'positive' plasma, which means that an electron beam was formed. In [10.59], the beam electron energy was 20 to 30 eV, the argon plasma density was $n = 5 \times 10^8\,\mathrm{cm}^{-3}$ and the neutral argon pressure was $p = 1\text{-}2 \times 10^{-4}\,\mathrm{mm\ Hg}$. The beam density was 5 to 10% of the plasma density (at the beam current of several amperes). A signal at a frequency close to the Langmuir frequency $f_p = \omega/2\pi = 200\,\mathrm{MHz}$ was fed to the input of the system by grids and a monochromatic Langmuir wave was then generated in the plasma. Experiments demonstrate that the space-time evolution of this wave depends critically on the amplitude of its electric field E_0. If the field strength is sufficiently low (but is above a certain threshold $E_0^2/16\pi nT \simeq 0.1\%$), the wave undergoes a parametric decay of the type of backscattering on ion-acoustic wave

$$k_0 = -k_0 + k_s \qquad k_s = 2k_0. \qquad (10.38')$$

where k_0, $-k_0$ and k_s are the wave vectors of the incident, scattered and ion-acoustic waves, respectively; the spatial period of this last wave is $\lambda = \pi/k_0$. If the field strength of the original Langmuir wave is increased, then the picture changes dramatically at

$$\frac{W_0}{nT} \equiv \frac{E_0^2}{16\pi nT} \gtrsim 0.3\%. \qquad (10.39)$$

Namely, the travelling ion-acoustic wave is replaced by a sequence of stationary density wells, at a period λ_s, into which the electric field of

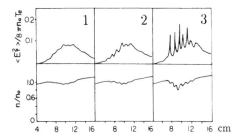

Figure 10.2 Spatial distributions of electric field energy and plasma density at three moments after the pump wave was switched on. Horizontal coordinate: distance z from the controlling grid creating the electron beam; $E_0 = 5\,\mathrm{V\,cm}^{-1}$ [10.59]. 1, $t = 10\,\mu s$; 2, $t = 20\,\mu s$; 3, $t = 30\,\mu s$.

Langmuir waves is 'captured': the maxima of the Langmuir wave field coincide with the minima of the plasma density (figure 10.2). As E_0 increases, the Langmuir waves localized in density wells are shown by the experiments to self-compress still further: the characteristic width of wave bunches decreases to at least $10r_D$, and the value of W/nT reaches 10–20%.

The increment of this process in the pulsed beam regime has been measured: the characteristic time at E_0 (three times the threshold value (10.39)) was of several tens of microseconds, that is, was equal to several periods of ion-acoustic oscillations ($T_s = 2\pi/k_s c_s$, where $c_s = (T_e/M)^{1/2}$ is the ion-acoustic velocity, T_e is the electron temperature and M is the argon ion mass). The authors were unable to follow further evolution of the waves after their self-compression into bunches narrower than about $10r_D$ and so could add nothing more to the description of their experiment (figure 10.2).

It is important to point out that the following relation was observed between the value of $W/nT \simeq (10\text{–}15)\%$ in wave bunches and their characteristic size $\Delta \simeq 10r_D$:

$$\frac{W}{nT} \simeq \left(\frac{\pi}{\Delta}r_D\right)^2 \simeq (\kappa r_D)^2 \qquad (10.40)$$

where $\kappa \simeq \pi/\Delta$ is the characteristic wave number.

The experiments of [10.73] had a similar physical meaning. They were conducted in a set-up not very different from that of [10.59] (the double plasma device type); the beam and plasma parameters were also approximately the same (figure 10.3). These experiments were specific in that prior to the injection of the beam, a high-amplitude ion-acoustic wave was passed through the system; it was generated at the input to the system by grids to which an AC potential at a frequency $f_s = 300\,\mathrm{kHz}$ was fed as a short pulse. The amplitude $\delta n/n$ of this wave was about 1%.

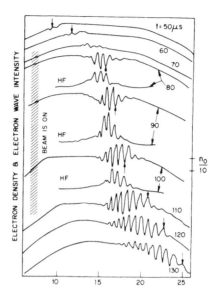

Figure 10.3   The same as in figure 10.2 under the conditions of [10.73]. The beam was on only during the interval from 70 to 100 μs. After the beam was turned off ($E \rightarrow 0$), only the ion-acoustic wave remained (110, 120 and 130 μs times). Arrows point to the fixed wave phase.

The experimental conditions were adjusted so as to meet the following relation between the wave numbers k_s of the ion-acoustic wave and k_p of the Langmuir wave excited by the beam in the plasma:

$$k_s = 2k_p \qquad (10.41)$$

where $k_s = \omega_s/c_s$ and $k_p = \omega_p/u$; $\omega_s = 2\pi f_s$ and $\omega_p = 2\pi f_p$ are the ion-acoustic and the plasma frequencies, respectively. This relation is equivalent to the condition

$$\frac{\omega_s}{c_s} \frac{u}{\omega_p} = 2 \qquad \text{or} \qquad \frac{2\pi f_s}{(T_e/M)^{1/2}} \frac{u}{(4\pi n e^2/m)^{1/2}} = 2. \qquad (10.42)$$

If the amplitude of the Langmuir beam–plasma wave is above a certain threshold, then a parametric backscattering-type decay of this wave (see (10.38)) is observed, as reported also in [10.59]. This process is greatly accelerated if there is an ion-acoustic wave in the system (with amplitude $\delta n/n \simeq 1\%$) and condition (10.42) is satisfied. Experiments demonstrate that three new phenomena are then observed: (1) the ion-acoustic wave amplitude increases by approximately an order of magnitude, (2) the electric field of the Langmuir wave is captured into the ion-acoustic density wells and (3) the propagation velocity v_p of the ion-acoustic wave ('loaded' with the captured Langmuir wave) is found to be significantly lower than the ion-acoustic velocity: in typical conditions, $v_p \simeq 0.6c_s$.

Bound waves are, therefore, excited in this particular case in contrast to experiments [10.59]; these bound waves were the Langmuir and the ion-acoustic waves; note that the latter wave does not degenerate to a sequence of standing stationary density wells but moves at a velocity of about one half of c_s. This difference is a result of the initial and boundary conditions formulated above.

The plasma density modulation amplitude $\delta n/n$ in the bound waves of this experiment reached about 20% and the relative energy density of the Langmuir waves captured into the density wells, moving at a velocity $v_p \simeq 0.6c_s$, was $W/nT \simeq 10\%$, that is

$$\frac{W}{nT} \simeq \frac{1}{2}\frac{\delta n}{n} \simeq \frac{1}{2}(k_s r_D)^2 \qquad (10.43)$$

(the coefficient $1/2$ appears because the wells are not stationary but move).

Therefore, the experiments of [10.59, 10.73] revealed the instability of the Langmuir waves with respect to the modulation with wavelength $\lambda_s = \lambda/2$; in the terminology of section 10.2, it was the short-wavelength modulational instability. It was mentioned already that this instability is 'indifferent' to the Lighthill condition (10.15) and thus develops quite freely in a low-energy electron beam: $W_1 \simeq 25\,\text{eV}$.

The structures described here are the so-called *cavitons*—non-linear Langmuir waves which are localized in the plasma density wells they themselves generate. At the same time, they constitute *regular* entities whose spatial period coincides with the original wavelength (or one half of it) and whose profile differs from a sine wave in a non-linear sharpening of the wave crest. Consequently, the structures observed are not yet 'lone' solitons but rather *cnoidal waves* [10.26]. The Langmuir solitons are described below.

The following experiments [10.77] were run with a *non-modulated* electron beam of relatively high energy: $W_1 \simeq 800\,\text{eV}$. The plasma density was $n_e \simeq 2 \times 10^9\,\text{cm}^{-3}$, the beam density was $n_1 \simeq (0.2\text{--}4)\%$ of n_e, the beam radius — 3 to 4 cm, the beam length — about 100 cm. Experiments demonstrated that beginning with a certain beam density, cavitons are generated which confirm the above description but are spatially less regular. These cavitons are not stationary: the electric field strength and the depth of the density well grow rapidly with time. The characteristic evolution period of the process is of the order of several thousand periods of Langmuir oscillations (i.e., of the order of 10 μs). A typical example of such structures is shown in figure 10.4 which plots the spatial distribution of the electric oscillation field (upper curve) and that of the plasma density (lower curve) at a time close to the saturation of instability when $E_0^2/(16\pi nT) \approx 1$ and $\delta n/n \simeq 0.6$.

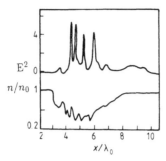

Figure 10.4   Bunches of Langmuir wave energy (E^2) (upper curve) and the corresponding plasma density 'wells'. The ratio of the beam density to the plasma density is 2%. The beam electron energy $W_1 = 800\,\text{eV}$ [10.77].

Neither of these structures are solitons, since they are highly non-stationary. They are known as *spikons* (see below).

The time evolution of these structures was studied in [10.78] by the same group of authors. The following spatial and temporal evolution of these structures was demonstrated (figures 10.5, 10.6). The electron beam first generates Langmuir waves in a relatively large region whose z extension (along the beam propagation direction) is about $500 r_D \simeq 10\,\text{cm} \simeq 3\lambda_0$ and whose radial extension is about $150 r_D \simeq 3\,\text{cm}$ (here λ_0 is the wavelength of the Langmuir wave generated by the beam). At subsequent times (while the beam is still on), the region of Langmuir waves undergoes self-compression and contracts along z by a factor of about 25 (down to a size of $\simeq 20 r_D \simeq \lambda_0/8$) and along r by about 3 to 4 (down to a size of $\simeq 40 r_D$). This phenomenon is in qualitative agreement with the theoretical concept of collapse of Langmuir waves (see [10.25, 10.80–10.82]). The minimal observed size of collapsed cavitons (spikons) is dictated by Landau damping of Langmuir waves [1.15, 1.16]. In this particular case, the damping is caused by the absorption of waves by above-thermal electrons which are created during the wave collapse; the energies of these electrons are greater than the temperature of most plasma electrons by approximately an order of magnitude [10.79, 10.83].

It was also shown that the maximum value E_S that the electric field of Langmuir waves reaches in time grows with increasing relative beam density $\alpha = n_1/n_e$. According to figure 10.7, $E_S^2 \propto (n_1/n_e)^{1.34}$.

Figure 10.8 plots the velocity distribution function of beam electrons at three stages of interaction with the plasma: before the excitation of Langmuir waves, in the course of excitation and after the non-linear saturation of the wave amplitude (see section 10.1). It illustrates an interesting result: the beam–plasma interaction gets considerably weaker at the last stage of the interaction: the plasma becomes more 'transparent' to the beam (a similar effect is mentioned in section 10.9).

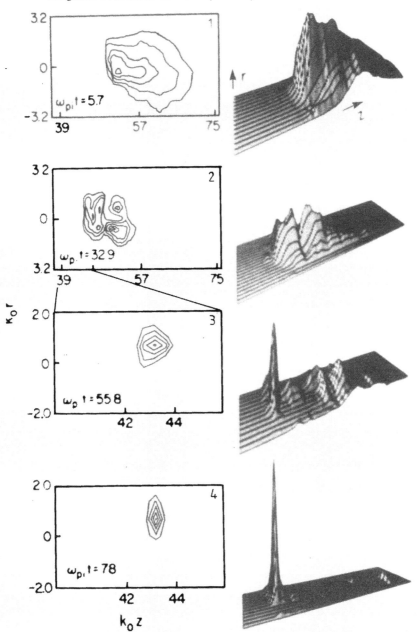

Figure 10.5 Two-dimensional cross sections and three-dimensional profiles of Langmuir wave intensity $E^2(r, z)$ at four moments of time: 1–4, $\omega_p t = 5.7, 32.9, 55.8,$ and 78, respectively. The outer contour corresponds to E^2 equal to 0.35 of the maximum value, and the other contours correspond to equal increases in E^2 [10.78]. Shown on the right are the three-dimensional distributions of the electric field in the waves.

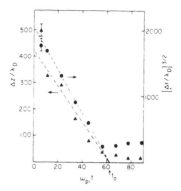

Figure 10.6   The rate of reduction of spatial dimensions of the Langmuir caviton. The axial sizes (triangles) are given in units of $\Delta z/r_D$ and the radial sizes, in units of $(\Delta r/r_D)^{3/2}$. The quantities Δr and Δz correspond to one half of wave intensity [10.78].

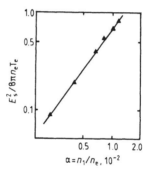

Figure 10.7   Maximum electric field of Langmuir waves as a function of the ratio of beam density to plasma density [10.79]: $E^2 \propto \alpha^{1.34}$.

The experiments demonstrated the above-mentioned phenomenon of 'plasma transparency' in the beam–plasma system during the evolution of the modulational instability.

The self-compression of Langmuir waves in a non-magnetized plasma was also observed in [10.84].

10.4   Langmuir Solitons in Magnetized Plasma

The cavitons described in the preceding section are three-dimensional wave packets which collapse with time; hence, they are not Langmuir solitons.

The creation of Langmuir solitons and an analysis of their properties were first achieved in experiments with magnetized plasma; their description is given below. It is necessary to point out first of all that

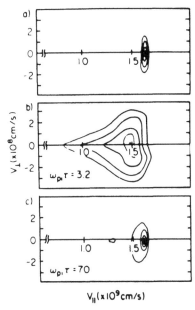

Figure 10.8   Contours of longitudinal and transverse velocity distribution function $f_{beam}(v_\parallel, v_\perp)$: (a) the initial time, $f_b(v_\parallel, v_\perp)$ at $\tau = 0$, $\Delta v_\parallel/v_0 = 1.6\%$ and $\Delta v_\perp/v_0 = 8.3\%$, (b) before wave collapse, $f_b(v_\parallel, v_\perp)$ at $\omega_{pi}t = 3.2$, when $\Delta v_\parallel/v_0 = 26\%$ and $\Delta v_\perp/v_0 = 22\%$ and (c) after collapse $f_b(v_\parallel, v_\perp)$ at $\omega_{pi}t = 70$, when $\Delta v_\parallel/v_0 = 4.4\%$ and $\Delta v_\perp/v_0 = 10.2\%$. The quantity ω_{pi} is the ion Langmuir frequency. The contours are traced for equal intervals of f_{beam}; the outer contour corresponds to 0.3 of the maximum value [10.78].

a strong external field is a factor of principal importance, which makes the system one-dimensional and thus forbids the (three-dimensional) collapse of the wave; hence, it stimulates the formation of (one-dimensional) solitons— see section 10.2.

Magnetized plasma experiments had the following characteristic features, as compared with the experiments described in the preceding section. First, they could be run both with the beam on and after the beam was turned off, while the non-magnetized plasma experiments described above were possible only when the beam was on. This difference was caused by the relatively high working gas (argon) pressure in the region of the beam in the former experiments: $p \simeq 10^{-4}$ mm Hg. The frequency of collisions between plasma electrons and gas atoms was therefore so high that Langmuir waves were damped out by collisions almost immediately after the 'pump beam' had been turned off. Consequently, the question about the behaviour of *free* packets of Langmuir waves in these experiments (with the pump beam turned off) remained unanswered. Neither could

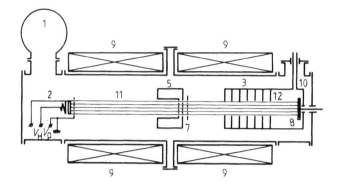

Figure 10.9   Experimental set-up for studying Langmuir
solitons in magnetized plasma [10.60.61].

these experiments give answers about the stability of the one-dimensional
Langmuir soliton. However, experiments with magnetized plasma used a
different method for plasma generation and the gas pressure in the beam
path was lower by 1.5–2 orders of magnitude. As a result, it did not set a
limit to the lifetime of Langmuir waves after the beam was turned off.
Secondly, the magnetized plasma experiments were run with a *moving
plasma* — in order to be able to detect solitons, which were stationary (or
ceased to move) with respect to the plasma, using stationary diagnostic
equipment.

Experiments on the generation of Langmuir solitons in magnetized
plasma were carried out in [10.60, 10.61, 10.74, 10.75].

The experimental set-up (see figure 10.9) included the following
elements: a 250 cm long solenoid, 9, creating an approximately
homogeneous magnetic field (in these experiments, $H = 2 \times 10^3$ Oe); a
vacuum volume evacuated by diffusion pump, 1; discharge chamber, 12
in which the plasma was generated by passing an electron beam from
electron gun, 2, through the gas (hydrogen) fed into the chamber; a
set of diaphragms which compose a so-called gas delay line, 3, 80 cm
in length, which removed from the plasma, for a time sufficient for the
experiment, the neutral gas travelling with it along the magnetic field at
a velocity of the order of c_s; plasma column 11 of 3 to 4 cm in diameter;
beam collector, 8; working space of length $L = 100$ cm with diagnostic
resonator, 5 (the working slit of the resonator is covered by a grid of
about 95% transparency); HF probe 7 and a receiver for recording Langmuir
oscillations. The plasma was generated in a discharge chamber which was
pulse-filled with hydrogen ionized by an electron beam pulse 10 to 20 μs long
at energy of 0.5 to 2 keV and current of tens or hundreds mA. The plasma
'leaked' from the discharge chamber along the magnetic field and reached
the working space some 50 μs after the beam was turned off; there the

plasma continued to propagate at (approximately) the velocity of ion sound $c_s \simeq 2 \times 10^6$ cm/s. The gas pressure in the working space (in the plasma flow regime) was about 3×10^{-6} mm Hg (it remained at this level for 10 ms after the plasma flowed through). The electron temperature (evaluated using the Langmuir probe) was $T_e \simeq 10$ eV, typical plasma density $n \simeq 3 \times 10^9$ cm^{-3}, $r_D \simeq 3 \times 10^{-2}$ cm. The electron beam used to pump Langmuir waves into the plasma was turned on independently of the beam which generated the plasma; also, it had independent parameters and a pulse length which was approximately equal to the plasma pulse length. The plasma with the parameters mentioned above was magnetized ($\omega_H \gg \omega_p$) and collisionless:

$$\tau_e = \{n_0 \langle \sigma v \rangle_0 + n \langle \sigma v \rangle_e\}^{-1} \simeq 10^{-4}\,\text{s}$$
$$\tau_s = L/c_s \simeq 0.5 \times 10^{-4}\,\text{s} \tag{10.44}$$

that is, the characteristic time of electron collisions was greater than the 'time of flight' of ion sound along the system,

$$\tau_e > \tau_s. \tag{10.45}$$

Here n_0 is the concentration of neutral hydrogen in the working space, v is the plasma electron velocity, σ_0 and σ_e are the effective scattering cross sections of plasma electrons in collisions with hydrogen molecules and charged plasma particles ($\sigma_0 = 5 \times 10^{-16}$ cm^2, $\sigma_e = 6 \times 10^{-13} T_e^{-2}$ cm^2, where T_e is measured in units of electron volts) [10.85, 10.86]. (Note, for comparison, that the argon pressure in the experiments [10.59, 10.73] was $p = (1\text{–}10) \times 10^{-4}$ mm Hg, that is, greater by about two orders of magnitude, while the sound velocity was lower (owing to the greater ion mass M); as a result, the inequality $\tau_e \ll \tau_s$ was satisfied, that is, the plasma regime was collisional in principle and the observation of solitons was impossible after pumping was turned off.) In the geometry described above, the plasma propagates towards the beam and therefore the beam electrons entering the plasma at the entrance to the working volume can be regarded as monoenergetic. The beam diameter in the region of the electron gun was 3 cm, and could be increased to 6 cm within this volume. The increase of the beam diameter was achieved by reducing the magnetic field strength in the working space by a factor of three, with the same field at the electron gun and the same cathode diameter. The beam pulse duration could reach 500 μs; the measurements were carried out both during the pulse and after it. The beam electron energy W_1 was 200–1000 eV, the current was $I = 50\text{–}80$ mA, the magnetic field strength was $H = 500$–1000 Oe. The plasma propagation velocity v in these experiments was, depending on specific geometry, from 4×10^5 cm/s to 2×10^6 cm/s, that is, it was within the interval $(0.2\text{–}1)c_s$.

The plasma density was measured and monitored by the resonator method; in this particular study, this method was realized as follows. As

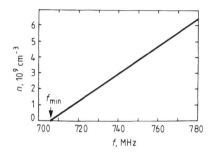

Figure 10.10   Natural frequency of the diagnostic resonator as a function of plasma density in the resonator gap.

the plasma travelled through the capacitor gap of the coaxial quarter-wave resonator, its resonant frequency changed in accordance with the changes in the dielectric permittivity of the medium in the gap; the dielectric permittivity of that part of the gap which was filled by the plasma (in the oscillation mode selected in our experiments for plasma diagnostics (see below)) was given by expression (4.2)

$$\varepsilon = 1 - (4\pi e^2/m\omega^2)n$$

where ω is the frequency of the diagnostic resonator ($\omega > \omega_p$).

If the resonator is pumped by an external generator at a certain 'vacuum' resonant frequency in the absence of the plasma, then the plasma should detune the resonator, by increasing its resonant frequency (figure 10.10). Consequently, it decreases the amplitude of field oscillations in the resonator, as measured by a diagnostic loop placed at the maximum of the magnetic component of the field (the detected signal from the loop is fed to the storage oscillograph). To restore the resonance at a given plasma density, the frequency of the supply generator must now be increased, and the higher the density, the greater the increase required. The resonator is calibrated using a special set-up (its resonant frequency is determined as a function of ε in the relevant part of the capacitive gap). The material used for this is Teflon which changes the gap dielectric constant by unity (permittivity of Teflon is $\varepsilon = 2$); the resulting change in the absolute dielectric constant is the same as would be obtained by filling the gap with the plasma at the limiting density $n = m\omega_p^2/4\pi e^2$ (which produces $\varepsilon = 0$). The plasma density is then readily found by measuring the resonant frequency of the resonator (i.e. the frequency of the supply generator at which the resonance is reached) when the plasma moves through the gap. It is not difficult, therefore, to find the spatial profile of the plasma or (which is equivalent) its time profile by varying the generator frequency and tracing the oscillograms of the (detected) signal of the diagnostic loop (figure 10.11). In this way it is possible, for example, to detect in the plasma a density well connected with the Langmuir soliton. The absolute

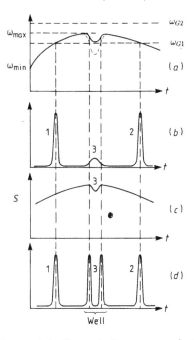

Figure 10.11 (a) Natural frequency of the diagnostic resonator as a function of time as the plasma column with a region of reduced density δn travels through the gap, and (b)–(d) the diagnostic loop signal, s, as a function of time: 1, the advancing front of the moving plasma; 2, the rear front; 3, the reduced density region.

measurements of the plasma density well are performed as follows. Assume that the oscillogram obtained is of the type shown in figure 10.11. Now the absolute density change δn in the well is found from the calibration curve of the natural resonator frequency as a function of plasma density in the resonator gap. The same plot gives the absolute plasma density n at the frequency of the supply generator, which corresponds to the maximum loop signal (peaks of pulses 1 and 2 in figure 10.11). The value given below is the depth of the well $\delta n/n$ measured by this procedure. The natural (vacuum) frequency of the diagnostic resonator was 720 MHz.

To clarify these statements, we turn to figure 10.11 which shows schematically how the resonant frequency ω_{res} of the diagnostic resonator changes with time from ω_{min} to ω_{max} as the plasma column with reduced density passes through the gap (ω_{min} corresponds to the absence of plasma, and ω_{max}, to the presence of plasma at its maximum density). If the generator frequency ω_G is in the interval $\omega_{\text{min}} < \omega_G < \omega_{\text{max}}$, then the shape of the signal measured by the resonator detector loop is similar to that shown in figure 10.11(b): the signal reaches a maximum at the points $\omega = \omega_G$, and the regions of reduced density correspond to a local increase of the loop signal. If, however, $\omega_G > \omega_{max}$, the regions of reduced density correspond to a decreased loop signal (see figure 10.11(c)). The sensitivity of this method is a function of the system's geometry and the resonator quality factor; wells of 2 to 3% relative depth could be reliably detected in the experiments.

An HF probe was used to indicate the relative electric field strength of Langmuir oscillations. The working surface of the probe was a grid of 4 cm in diameter, with 2×2 mm meshes of 0.1 mm diameter tungsten wire. The probe was placed in the plasma stream so as to have the probe plane perpendicular to the stream velocity. The probe signal was measured by two independent methods. In the first one, the signal was fed via a coaxial cable to the 75 ohm input of a selective tunable receiver; after detection, the signal was fed from the receiver output to the storage oscillograph and photographed. The receivers made it possible to single out a fixed frequency in the 250–1000 MHz range, with a wave band of 0.8 MHz. In the second method, the signal was fed to a detector which operated in the 50–10 000 MHz frequency range, and then to a broad-band amplifier; I will refer to such a signal, which includes HF oscillations at practically all frequencies (not only the frequency we select) as the *integral signal*.

No measurements of the absolute values of the electric field in Langmuir waves were carried out in this work. In principle, they are not absolutely necessary because the quantity $E_0^2/16\pi$ in a soliton is determined from the depth of the density well $\delta n/n$ (see relations (10.7) and (10.27)). In the experiments [10.59, 10.73] described above, where $H = 0$, a transverse electron beam was used as a probe for measuring the absolute electric field strength. This method could not be used in this particular work because of the presence of strong magnetic field.

The first result of the experiment on generating Langmuir waves in a magnetized plasma was as follows. When the pump beam current exceeds a certain threshold (about 3 mA at the beam electron energy about 100 eV), the Langmuir oscillations generated by the beam in the plasma take the form of relatively rare but intense electric field pulses at a frequency of 500 MHz ($\simeq f_p$). At the same time, a clearly pronounced plasma density modulation arises: density wells appear at the points of localization of wave bunches, with the electron concentration reduced by at most 5–10% in comparison with the density of the surrounding plasma. This can be seen in figure 10.12, where the upper oscillographic trace (CRT ray deflected downwards) is the envelope of the electric field of the Langmuir wave (at a frequency $f = 500$ MHz $\simeq f_p$), and the lower trace (CRT ray deflected upwards) is the oscillogram of the diagnostic resonator (see [10.61]).

The width Δ_E of the wave bunches localized at the plasma density wells is smaller, the higher their amplitude (or depth of the well). If $\delta n/n = 0.05$–0.1, this width is definitely below 1 cm, that is, below $(25$–$30)r_D$; this estimate includes the time resolution (1 μs) of the HF receiver, and is thus

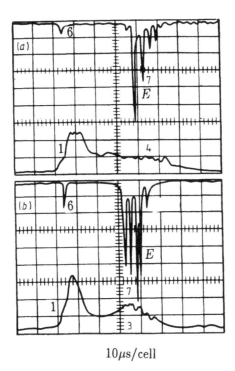

$10\mu s/\text{cell}$

Figure 10.12 Oscillograms of the electric field E of Langmuir waves (CRT ray deflected downwards) and of plasma density (CRT ray deflected upwards) (plotted in arbitrary units). 1, 3, 4, plasma density; 6, 7, electric field bunches.

an upper-bound estimate.

This picture of Langmuir bunches and the related density wells changes substantially when the diagnostic equipment is moved along the working space. While a well-developed modulation of plasma density and a fairly large number of Langmuir bunches are observed at relatively large distances ($\Delta z = 30$ cm) from the starting point of the working volume (see figure 10.12), the density modulation is much less developed and only rare Langmuir bunches are observed at short distances ($\Delta z = 10$ cm). This means that the characteristic length of evolution of wave bunches is of the order of 10 cm, that is, the characteristic time is about 10 μs (several thousand Langmuir periods).

Measurements demonstrated that wave bunches localized in plasma density wells move together with the plasma, that is, they are, in the first approximation, stationary relative to the plasma itself.

Figure 10.13(a) plots the delay time τ of the arrival of an HF field bunch

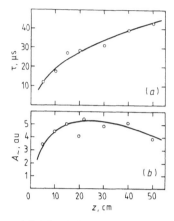

Figure 10.13 (*a*) The delay time τ of the arrival of an HF field bunch to the probe as a function of probe displacement along the working volume and (*b*) electric field amplitude in a bunch as a function of the propagation distance, $z \equiv \Delta z$.

to the probe as a function of probe displacement along the working volume†.

We see that wave bunches travel freely in the set-up over at least 50 cm, in a time of about 40 μs. In other words, the lifetime of field bunches equals at least 2×10^4 periods of electron plasma oscillations. Bunches observed at larger distances from the starting point of the working volume are typically of greater absolute value and narrower than those at short distances. The amplitude $A_\sim$ of the electric field in a bunch as a function of the distance travelled by the bunch is plotted in figure 10.13(*b*) (in arbitrary units).

Bunches of Langmuir waves observed in the experiments described above are essentially non-linear entities. Indeed, as shown in section 10.2, a linear packet of the size observed ($\sim 25r_D \simeq 1$ cm) would spread out already at a distance equal to the initial bunch width, while the actual bunches travel over distances tens of times greater without appreciable spreading. In view of all the attributes observed, the Langmuir wave bunches localized in a plasma density well can be identified as Langmuir solitons.

This work was thus the first in which solitons were observed to form as a result of compression of non-linear Langmuir waves.

The wave self-compression has the characteristics of instability. The time of evolution of this instability (several microseconds) is quite sufficient, from the theoretical standpoint, for soliton formation (see section 10.2). The sizes of the observed self-compressed wave packets (25–$30\,r_D$ at the most) are fairly close to the theoretical sizes of Langmuir solitons. Indeed, relation (10.27) implies that with the observed width $\delta n/n \simeq 5\%$ of the density well, the soliton width must be $\Delta_E \simeq 35r_D$, while the experiments point to $\Delta_E \simeq 25r_D$. Solitons are relatively stable: their 'lifetime' is not

† These measurements were conducted after the beam was turned off.

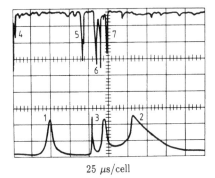

25 μs/cell

Figure 10.14   Same as in figure 10.12. 1, 2, the advancing and the rear fronts of the propagating plasma; 3, density well (produced artificially in this experiment), 6–7, wave bunches at the middle of the well and at the rear front.

less than 30–40 μs (tens of thousands of Langmuir oscillation periods), and their free path (together with the moving plasma) is not less than several tens of centimetres.

It is of interest to compare the size of the solitons observed with the wavelength λ of Langmuir waves excited by the beam in the plasma. Under the conditions of figure 10.12, we find: $\lambda = u/f_\mathrm{p} = 6 \times 10^8/5 \times 10^8 = 1.2$ cm, $\Delta_E \simeq 0.7$ cm, that is,

$$\Delta_E \lesssim \lambda. \tag{10.46}$$

This situation is typical of a threshold of short-wavelength modulational instability.

The difference between the experiments of this paper and those of [10.59, 10.73], which stems from the collisionless regime in the plasma, is additionally demonstrated by the oscillograms of figure 10.14 which were obtained in the regime of short-duration pump beam (about 3 μs, in contrast to all the other data of this section, where the pulse duration was 200 μs). We see that solitons are observed with a considerable delay after the beam has been turned off: the delay reaches 35 μs already at a distance of 20 cm from the starting point of the working volume, while in [10.59, 10.73] solitons vanished owing to collisional damping, virtually instantly after the removal of pumping.

With the plasma propagation velocity known, we can easily transform the time scale of plasma density variation into a spatial scale. Correspondingly, the signal of the diagnostic resonator loop can be called either the *temporal* or the *spatial plasma density profile*. The following feature needs to be borne in mind when this profile is compared with an oscillogram of the HF probe (used as the indicator of electric field bunches of Langmuir waves). The diagnostic resonator (see figure 10.9) is displaced with respect to the HF probe by 2.5 cm in the direction of plasma propagation. Therefore, if a soliton is, say, stationary with respect to the

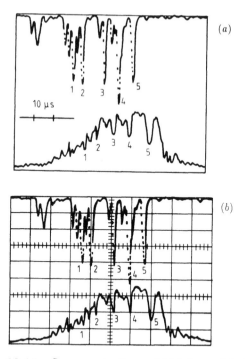

Figure 10.15   Same as in figure 10.12 but with different beam and plasma regimes: plasma density $n = 1.8 \times 10^9$ cm^{-3}, $W_1 = 900$ eV, $f = 350$ MHz. Figure (b) differs from (a) in a shift of the oscillogram of electric field by 5 μs to the right.

plasma, it is detected by the resonator (as a density well) later than by the HF probe (as a field bunch). For this reason, and also owing to inertia of the HF field receivers (~ 1 μs), the observed delay of the density well relative to the field bunches was about 5–6 μs.

The subsequent [10.61] as well as the earlier experiments [10.60] demonstrated that the Langmuir waves generated by the beam in the plasma constitute a set of solitons, that is, bunches of Langmuir-wave electric field connected with deep density wells ($\delta n/n = (3\text{–}30) \times 10^{-2}$). Solitons are localized in the plasma, that is, are (in a first approximation) stationary relative to it: if they move in the plasma, their velocities are much lower than the ion-sound velocity c_s. These conclusions will now be illustrated by several oscillograms. Typical oscillograms of the signal at the plasma density detector as a function of time and those of the envelope of the Langmuir wave electric field are shown in figures 10.15, 10.16 (the values of plasma density given in the captions to these figures refer to the time at which the diagnostic resonator signal reaches a maximum). The following features appear to be especially significant.

The oscillations are indeed of Langmuir type: their frequency f is given

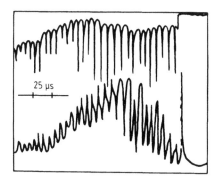

Figure 10.16 Oscillograms of the integral HF probe (upper curve) and of the diagnostic resonator signal, at plasma density $n = 1 \times 10^9$ cm^{-3} and $W_1 = 900$ eV. The intensity maximum in the oscillation spectrum lies in the region of 300–350 MHz. The relative density reduction in the deepest wells reaches $\delta n/n = 0.3$. The signal cutoff at the ends of the oscillograms is caused by the beam turn-off [10.61].

by the formula $f \simeq 10^4 \, n^{1/2}$.

Langmuir waves form wave bunches. The characteristic period of these bunches (several microseconds) is greater than the Langmuir oscillation period by a factor of approximately 10^3. As a result, these field bunches are in fact 'Langmuir-wave filled' solitary waves.

The number of field bunches coincides roughly (see figure 10.15) or exactly (see figure 10.16) with the number of plasma density wells.

If we assume that field bunches and density wells move together with the plasma, and take into account the above-mentioned delay of detecting wells as compared to detecting field bunches (5–6 µs), we can conclude that under these experimental conditions field bunches are usually localized in plasma density wells. A closer analysis of several figures will help to clarify this conclusion. If the lower ('retarded') oscillogram in figure 10.15(a) is shifted to the left by approximately 6 µs (as shown in figure 10.15(b)), the times of the field pulses marked by numbers 1–5 very closely coincide with those of the corresponding density wells. This correspondence of field bunches and density wells is seen especially clearly in figure 10.16 where we see that all field bunches, without a single exception, are localized in their corresponding density wells. Therefore, the most reliable interpretation of the experimental data presented above is this: the wave entities in question are Langmuir solitons moving together with the plasma (in the plasma reference frame they are at rest).

Conditions are possible under which the integral signal of the Langmuir wave field indicator has the same time modulation as the signal at the selected frequency (in the case of figure 10.16, at the frequency $f \simeq f_p \approx$ 300 MHz). This means that waves at all frequencies of the HF spectrum

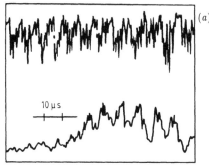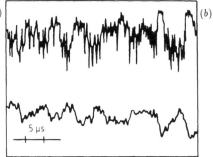

Figure 10.17   Examples of modulation of the integral signal of the HF probe (upper trace) and of the signal of the diagnostic resonator (lower trace), W_1 = 900–1000 eV; (a) $n = 4 \times 10^9$ cm^{-3}, (b) $n = 4.5 \times 10^9$ cm^{-3} [10.61].

are localized in density wells and are virtually absent outside these wells. Under these conditions, the modulational instability of Langmuir waves is seen especially clearly.

It must be indicated that two patterns of wave modulation can be observed depending on experimental conditions: in one case there is one soliton per (each) modulation wavelength (see figure 10.16) and in the other, whole 'stacks' of solitons fall within one modulation wavelength (see figure 10.17). Not only wave bunches but also density wells are observed to form such stacks (see figure 10.17). The time period of the envelope of these stacks varies from 5 to 20 μs, that is, for the plasma propagation rate $v = 0.4 \times 10^6$ cm/s, the spatial period varies from 2 to 8 cm.

When field bunches can be put in correspondence with certain density wells, it is invariably found that higher-intensity bunches correspond to deeper and narrower wells (see, e.g., figures 10.15 and 10.16).

The depth of plasma density modulation in the wells (depth of the wells) lies in the range from 10 to 30%. Examples of the deepest wells are shown in figures 10.15 and 10.16 where $\delta n/n$ equals 10% and 30%, respectively. If, however, $\delta n/n \simeq 1$–2%, a well cannot be detected against the background of the inherent plasma density fluctuations.

The width Δ_n of density wells (the length of the well pulse times the plasma propagation velocity $v = 0.4 \times 10^6$ cm/s) is usually of the order of 0.5–1 cm. Examples of the narrowest wells whose width Δ_n is below ~ 0.3 cm (i.e. $\sim 10r_D$) are shown in figure 10.15 (pulses 3 and 4). It is important to mention that the measured width of the narrowest wells is practically equal to the gap width in the diagnostic resonator (0.25 cm, at the grid mesh width 0.2 to 0.25 cm). This means that the true well width may be even smaller than the estimate 0.3 cm given here.

Under specific (although fairly widespread) experimental conditions,

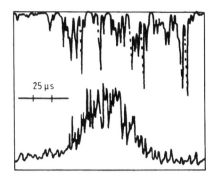

Figure 10.18 Oscillograms of the same signals as in figure 10.12, for $n = 2.2 \times 10^9$ cm^{-3}, $f = 550$ MHz and $W_1 = 900$ eV. Each sharp rise of the signal in the right-hand side of the lower oscillogram indicates a well, and each drop in the signal indicates a plasma density bunch. The situation in the left-hand side of this oscillogram is reversed: a rise in the signal reveals a plasma bunch while a drop in the signal signifies a density well [10.61].

there may be no detailed correlation between field bunches and density wells, even though the general type of field oscillations and density variations is almost the same for both (figure 10.18). Such cases can probably be regarded as examples of *soliton turbulence* [10.23–10.25, 10.64].

The period of modulation τ_M of Langmuir waves and the corresponding modulation wavelength $\lambda_M = v\tau_M$ ($v \simeq 0.4 \times 10^6$ cm/s) grow with increasing velocity of beam electrons. This dependence is shown in figure 10.19 whose data represent the results of averaging over a large number of beam pulses for each value of the beam-electron energy W_1. The same figure shows the wavelength λ of the linear Langmuir wave generated in the plasma by the beam:

$$\lambda = u/f_p \qquad (10.47)$$

where u is the beam-electron velocity, as a function of electron energy W_1 ($u = 6 \times 10^7 W_1^{1/2}$ cm s^{-1}, W_1 is given in electron volts). We see that λ_M and λ are not very different both numerically and in the way they depend on W_1.

Two regimes were observed in the experiments: with $\lambda_M = \lambda/2$ in one (as in figures 10.16 and 10.17) and with $\lambda_M = \lambda$ in the other [10.61].

Generally, deviations of the plasma density from the average value take the form both of density wells (described above) and of plasma bunches. The value of $\delta n/n$ in bunches may reach 10–30%.

The same experiments have demonstrated that the modulational instability of Langmuir waves — with clearly pronounced field bunches and the corresponding density wells (which stand out against the inherent

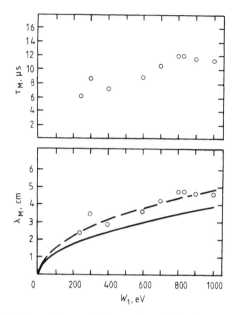

Figure 10.19   The period τ_M and modulation wavelength λ_M of electric field pulses of non-linear Langmuir waves, and the wavelength (solid curve) as functions of electron energy of the 'pumping beam' [10.61].

plasma noise)—is observed beginning with a certain threshold of oscillation energy, and that this threshold has a definite tendency to get lower as the beam electron energy W_1 is increased.

In order to analyse the oscillation spectrum, the HF probe signal was fed to two receivers connected in parallel and tuned to different frequencies. The measurements then demonstrated that if the difference Δf between receiver frequencies at the Langmuir frequency $f_p \simeq 500$ MHz was below 5 MHz, the oscillograms of the outputs of the two receivers were very nearly identical. For $\Delta f = 15$ MHz, the pulses recorded by the two receivers coincided in approximately 50% of events; for $\Delta f = 25$ MHz, the pulses coincided in relatively rare cases. In fact, it happened sometimes that oscillograms at $\Delta f = 50$ MHz were found to be quite similar. Measurements of this kind made it possible to come to the conclusion that the width of the frequency spectrum of Langmuir waves in the regime of soliton formation is $\Delta f = 15$ MHz, for the main frequency $f \simeq f_p \simeq 500$ MHz.

As to the 'free path' of solitons, figure 10.13 shows that it corresponds to a soliton lifetime of at least 100 μs and is thus at least longer than 50 cm. This result was also obtained in the independent experiments, described above, on the observation of Langmuir solitons after the beam pumping of waves was turned off (i.e. the beam was turned off); the Langmuir soliton lifetime in these experiments also exceeded several tens of thousands of

Langmuir oscillation periods (see figure 10.14).

We will now conduct a comparison of the observed properties of Langmuir solitons with the predictions of the theory, along the following lines:

(1) relation of the longitudinal size Δ (width) of a soliton to the depth $\delta n/n$ of the density well;

(2) relation of the characteristic scale of the modulational instability of Langmuir waves (modulation length λ_M) to the beam and plasma parameters;

(3) threshold of the observed modulational instability and its dependence on the beam electron energy;

(4) frequency spectrum of oscillations in the Langmuir soliton, and

(5) relation of the observed degree of self-compression of the Langmuir wave to the estimate of the 'coefficient of amplification' of its electric field.

Let us now consider the experimental data in the order given above.

(1) For comparisons of the soliton depth and width, turn first of all to figure 10.15. Under the experimental conditions of this figure (pulse 3), the soliton is in a density well $\delta n/n = 0.1$ deep and $\Delta_n = 0.25$ cm $\simeq 6r_D$ wide. At the same time, formula (10.28) implies that the soliton width is determined unambiguously by the depth δn of the density well:

$$\Delta_{n,\text{theor}} \simeq r_D \left(\frac{30n}{\delta n}\right)^{1/2} = r_D\left(\frac{30nT}{W}\right)^{1/2}.$$

Hence, the soliton width Δ_n under the conditions of figure 10.15 must be $\Delta_{\text{theor}} \simeq 18r_D$. This gives $\Delta_{\text{exper}}/\Delta_{\text{theor}} \approx 1/3$. In view of the current understanding of the physics of solitons, this ratio of the experimental and theoretical values of the longitudinal soliton size should be regarded as a satisfactory agreement rather than as a considerable discrepancy. The proportionality of the well depth and the field energy density, predicted by the theory, is also in a qualitative agreement with the experiment; this is also seen, for example, in figure 10.16. The relative depth of the deepest density wells, $\delta n/n \simeq 0.3$, is three times greater, and the plasma density $n = 1 \times 10^9$ cm^{-3} is three times lower, than in the conditions of figure 10.15. At the same time, the width of the highest-intensity solitons (in cm) is almost the same, and less by approximately a factor of $\sqrt{3}$ in Debye radii. This also agrees with formula (10.27). Under the conditions of figure 10.16, $\Delta_n = (5-6)r_D$. With the soliton width being so small, the theory predicts the soliton amplitude to be limited by dissipative processes (*Landau damping*) [1.15, 1.16].

Note that the larger width of the electric field pulse 1 as compared with the width of the well pulse 1 does not point to a contradiction with (10.27), which implies $\Delta_n \simeq \frac{2}{3}\Delta_E$.

(2) The experimental result shown, for instance, in figure 10.19 demonstrates for the length of modulation of non-linear Langmuir waves that $\lambda_M \simeq \lambda$ or $\lambda_M \simeq \lambda/2$ (see (10.36) and (10.47)). This means that the original Langmuir wave is an envelope of a shorter-wavelength modulation of the field due to the modulational instability; in other words, the original wave is found to be the envelope of solitons. This picture of non-linear waves also agrees with the theory [10.62]: it is characteristic of the short-wavelength version of the modulational instability treated in section 10.2. This version of the modulational instability is likely to develop at a sufficiently high density of oscillation energy,

$$\frac{W}{nT} \simeq \frac{\delta n}{n} \gtrsim (kr_D)^2$$

where k is the wave number of the oscillations modulated by the original Langmuir wave of wavelength λ (i.e. $\lambda > 2\pi/k$).

The situation corresponding to figure 10.17 can be characterized by the term 'soliton envelope', in contrast to the term 'envelope soliton' whose physical meaning was explained in section 10.2.

It was shown above that in the first approximation the observed Langmuir solitons are stationary with respect to the plasma, that is, they constitute bunches of standing non-linear waves in the reference frame of the moving plasma. The standing-waves picture takes the simplest form (see figure 10.16) when the modulation wavelength λ_M equals one half of the original wavelength, that is, it coincides with the spatial period of the wave intensity; i.e. with $\lambda/2$. In this case the wave forms in the plasma a regular sequence of density wells with a period $\lambda/2 = u/2f_p$ and gets localized in them, thereby creating a quasistable 'soliton lattice'. This self-compression of non-linear Langmuir waves is seen, for instance, in figure 10.16, which shows that there is virtually zero HF electric field between the arising narrow wave bunches. A more complex pattern is demonstrated in figure 10.17: the non-linear wave splits on a $\lambda/2$-long segment into several solitons of size $\Delta \approx (7-8)r_D$. Note that the process of formation of a standing non-linear wave (namely, of wavelength half that in the original pump wave) was observed experimentally [10.59] in strongly collisional non-magnetized plasma (and in computer simulations [10.87]). The pattern of further turbulence of the standing wave was calculated in detail in [10.62].

(3) The threshold (critical) energy density of Langmuir waves, beginning with which short-wave modulational instability arises, is determined, according to the theory of section 10.2, by formula (10.31):

$$\left(\frac{W}{nT}\right)_c \simeq (kr_D)^2.$$

It has been pointed out already that

$$k^2 > \left(\frac{2\pi}{\lambda}\right)^2 = \frac{4\pi^2 f_{\mathrm{p}}^2}{u^2} = \frac{2\pi n e^2}{W_1}$$

so that $(k r_{\mathrm{D}})^2 > T_e/2W_1$, which implies that

$$\left(\frac{W}{nT}\right)_c \gtrsim \frac{T_e}{2W_1}. \tag{10.48}$$

For example, if $W_1 = 800$ eV and $T_e = 10$ eV, the instability threshold is

$$\left(\frac{W}{nT}\right)_c \gtrsim 0.6 \times 10^{-2}. \tag{10.49}$$

Relation (10.48) shows that the threshold of short-wavelength modulational instability is determined to a large extent by the electron energy in the beam exciting the waves.

The experiments [10.61] have accordingly demonstrated that the modulational instability at relatively low beam electron energy ($W_1 = 400$ eV) lies relatively close to the excitation threshold, so that a sufficiently deep density well, clearly identified against the plasma noise background, cannot be formed. As the beam energy is increased to 800 eV, the data of [10.61] show that formula (10.48) indeed describes clearly pronounced wells in which are located the field bunches that initiated these wells. The depth $\delta n/n$ of these wells is $(4\text{--}5) \times 10^{-2}$, that is, it considerably exceeds the threshold given by (10.49). We can thus assume that the instability threshold was considerably exceeded in this case and that the wells became substantially deeper. They led to the formation of solitons whose width can be estimated as about $0.8\,\mathrm{cm}$ ($\simeq 20 r_{\mathrm{D}}$). According to the theory (see section 10.2), the characteristic wave number of these solitons is $k_0 \simeq 2/\Delta_n \simeq 1/10 r_{\mathrm{D}}$, that is, $(k_0 r_{\mathrm{D}})^2 \simeq 10^{-2}$, and according to (10.27), the field energy and well depth must now be $W/nT \simeq \delta n/n \simeq 6(k_0 r_{\mathrm{D}})^2 = 6 \times 10^{-2}$, while the experiment gave $\delta n/n = (4\text{--}5) \times 10^{-2}$. This comparison shows that there is a fairly good agreement between the theoretical and the experimental values of the threshold.

Let us now compare the experimental data and the theoretical relation (10.37) derived by synthesizing the theories of two instabilities: the beam instability and the modulational instability. According to (10.37) and (10.38), high-amplitude solitons can be formed of Langmuir waves only if the electron beam energy is greater than $250\,\mathrm{eV}$ (at $T_e = 10$ eV, as in [10.61]) or greater than $25\,\mathrm{eV}$ (at $T_e = 1$ eV, as in [10.59]). The experimental data shown in this and the preceding sections demonstrate that high-amplitude solitons indeed appear in roughly that energy range of beam electrons which is predicted by the theory. These data can thus be regarded as confirming the theory of the beam and modulational instabilities.

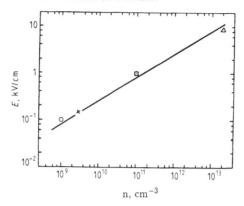

Figure 10.20   Electric field in bunches of Langmuir waves as a function of plasma density: ○ [10.59], × [10.60, 10.61], □ [10.76], △ [5.44].

The quantitative agreement of the theory and experiment allows the use of the theoretical relation

$$\frac{E_0^2}{16\pi} = \frac{\delta n}{n} nT \tag{10.50}$$

for the evaluation of the electric field strength in the Langmuir solitons observed. If $\delta n/n = 0.3$ and $n = 1 \times 10^9$ cm^{-3} (the conditions of figure 10.16) or if $\Delta n/n = 0.1$ and $n = 3 \times 10^9$ cm^{-3} (the conditions of figure 10.15), formula (10.50) gives, for $T_e = 10$ eV,

$$E_0 \simeq 150 \text{ V cm}^{-1}. \tag{10.51}$$

The following interesting fact must be mentioned in this connection. The electric field of Langmuir waves proved to be experimentally measurable. This was possible either because there was no magnetic field (so that a diagnostic electron beam could be used [10.59]) or because the field amplitude was high (so that spectroscopic methods were practicable [5.44]). The measurements showed, in accordance with (10.50), that the electric fields of Langmuir waves grew substantially with increasing plasma density; in the case of HF pumping, for example, the field reached $\sim 10^3$ V cm^{-1} at $n = 1 \times 10^{11}$ cm^{-3} [10.76] and could even reach $\sim 10^4$ V cm^{-1} at $n = 1 \times 10^{13}$ cm^{-3} [5.44] (in these experiments, the value of $\delta n/n$ was 0.05–0.1). The relationship $E \propto n^{1/2}$ was thus observed when n was varied from 10^9 to 10^{13} cm^{-3} (see figure 10.20). We can assume, therefore, that fields with strengths of the order of tens of megavolts per centimetre can be expected to arise in a denser plasma (e.g. a laser plasma where $n = 10^{19}$ cm^{-3} or greater). Of great interest in this connection is a result of numerical calculations in [7.23], which implies that the collapse of waves in a high-intensity quasineutral relativistic electron beam generates fields of strength $E_0 \simeq 20$ MV cm^{-1} (!). Such fields cannot but dominate the acceleration of charged particles in a beam [10.88].

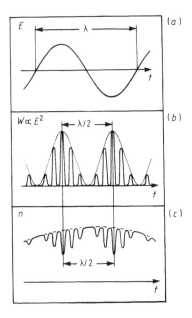

Figure 10.21 Spatial profile of the standing Langmuir wave (schematic representation): (*a*) field strength E, (*b*) field energy density $W \propto E^2$, (*c*) plasma density. The self-compression of waves and their fragmentation into several solitons occurs on a $\lambda/2$-long segment. This schematic corresponds to oscillograms of the type shown in figure 10.16.

(4) The frequency spectrum of non-linear Langmuir waves observed in modulational instability is found to be sufficiently narrow. The characteristic bandwidth of this spectrum was $\Delta f_{\mathrm{non}} = 10\text{--}15$ MHz $\simeq f_i$ where $f_i = f_p(m/M)^{1/2}$ is the Langmuir frequency of molecular hydrogen ions at the plasma density typical of our experiments, $n = 3 \times 10^9$ cm^{-3}, and M is the ion mass.

(5) As we see from figures 10.19, 10.21, a Langmuir soliton under the conditions of figure 10.16 concentrates the wave energy which was originally contained in a volume of length λ_M between $\lambda/2$ and λ; this means that the wave energy density W grows by a factor of approximately λ/Δ as compared with the original value W_0. For instance, if $\lambda = 4$ cm (see figure 10.19) and $\Delta = 0.2$ cm $\simeq 5r_{\mathrm{D}}$, then $W \simeq 20W_0$. If we assume $W_0/nT \simeq (1\text{--}2)\%$ (this seems very reasonable), we obtain $W/nT \simeq (20\text{--}40)\%$, which is in good agreement with the observed depression of plasma density: $\delta n/n \simeq 1/3$ (figure 10.16).

We can thus summarize the above material in the following conclusions.

1. As a first approximation, we find no significant discrepancies between

the experimental data and the theory, either qualitatively or in quantitative evaluations.

2. Langmuir solitons in a magnetized, collisionless plasma are very realistic, sufficiently stable and possess parameters and properties quite close to those predicted by the theory.

This result is far from trivial: when this work was at its initial stage, many theorists were very sceptical about the possibility of creating quasistable Langmuir solitons in radially confined magnetized plasma. Thus there was an apprehensive argument that 'almost Langmuir' waves at the frequencies

$$\omega = \omega_{\mathrm{p}}\frac{k_z}{k} = \omega_{\mathrm{p}}\frac{k_z}{(k_z^2 + k_\perp^2)^{1/2}} \qquad (10.52)$$

(i.e., 'slightly oblique' Langmuir waves) would propagate out tranversely from the plasma column at the group velocity $\partial\omega/\partial k_\perp \gg c_s$. However, this apprehension was not borne out: were it correct, the solitons, described above, with large free path would be unobservable (this conclusion holds even better for the phenomena discussed in the subsequent section). The absence of any appreciable transverse escape of waves from the plasma column is explained by the following natural factor: the formation of standing waves that do not propagate transversely.

10.5   Oblique Langmuir Solitons in Magnetized Plasma

This section treats not Langmuir waves but electron waves in a magnetized plasma in limited-diameter plasma column. The frequencies of these waves are below f_{p}. The dispersion of these waves (known as Trivelpiece–Gould waves) is given by relation (10.52) and figure 10.22†. We can refer to these waves as *oblique Langmuir waves* in confined plasma. I will discuss here experiments with solitons on the branch (10.52). The transverse wave number $k_\perp$ of these waves is determined by the plasma column radius; for instance, if the column is well separated from the walls, then $k_\perp \simeq 1/a$; among all the transverse oscillation modes, this (main) mode of cylindrically symmetric oscillations plays a decisive role in the properties of the waves discussed here. The problem of modulational instability of these waves is of interest also because they do not satisfy the Lighthill criterion (see section 10.2) and it is possible to test to what extent this factor affects the feasibility of soliton formation.

An experimental investigation of oblique Langmuir solitons was carried out in [10.74, 10.75], which extended the experiments of [10.60, 10.61]; the work was performed on the same experimental set-up. These experimental

† See also section 10.6.

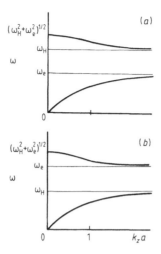

Figure 10.22 The frequency of electron oscillations of radially confined plasma column in longitudinal magnetic field as a function of wave number k_z and column radius a; (a) $\omega_p < \omega_H$, (b) $\omega_p > \omega_H$.

runs were conducted after the electron beam had been turned off for a considerable length of time, that is, in the phase of beam plasma afterglow. As in [10.60, 10.61], the plasma was collisionless. In order to 'get to the branch' (10.52), the plasma column diameter $2a$ was chosen to be 3 cm, that is, only half of the column diameter in [10.61]. In this case the column diameter is less than wavelength $\lambda = u/f$ of electron oscillations generated by the beam in the plasma (see below) and the oscillations observed correspond to the 'sought' mode (10.52). The general conditions and the methodology of the experiment were those of [10.61]: plasma density $n = (3\text{--}4) \times 10^9$ cm^{-3}, longitudinal magnetic field $H = 2 \times 10^3$ Oe, electron temperature $T_e = (10\text{--}20)$ eV, residual gas (hydrogen) pressure $p \simeq 5 \times 10^{-6}$ mm Hg, length of plasma column about 200 cm, plasma propagation velocity $V = (2\text{--}6) \times 10^6$ cm s^{-1}. The typical beam parameters were: electron energy $W_1 = (2\text{--}2.5)$ keV, current $I = (2\text{--}2.5)$ A, beam density $n_1 = (0.1\text{--}0.2)n$, beam pulse length about 20 μs. Under these conditions, the plasma frequency was $f_p = (500\text{--}600)$ MHz, beam electron velocity $u = (2.5\text{--}3) \times 10^9$ cm s^{-1}, wavelength of the Langmuir wave excited by the beam in the plasma $\lambda = u/f_p \simeq 5$ cm $> 2a$. The plasma propagation velocity was measured by the time shift of ion saturation currents to grid probes (with 95% transparency) placed at different points of the plasma column. The wave propagation velocity was measured by these same probes which were connected to two independent oscillation receivers tuned to the same frequency. This method of measurements made it possible to observe

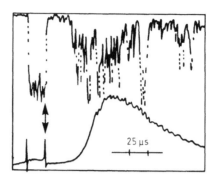

Figure 10.23 Oscillograms of plasma density indicators (lower trace, beam deflected upwards) and of the envelope of the electric field amplitude of HF waves (upper trace, beam deflected downwards). The indicators are located at the end of the plasma column at a distance of about 140 cm from the discharge chamber of the plasma source. As in all other oscillograms of this work, density wells correspond to downward deflections on the lower oscillogram; the arrow points to the time of turning off the pumping electron beam [10.74, 10.75].

the wave evolution in one 'shot' of the beam. The electric field indicators were, as in [10.60, 10.61], HF grid probes whose planes were perpendicular to the magnetic field. Therefore, the probes singled out from among the waves existing in the plasma column only those in which the electric field vector had a component parallel to the magnetic field. Correspondingly, either purely longitudinal (Langmuir) waves or oblique Langmuir waves could be detected among all HF waves propagating along the magnetic field.

I will now outline the main results of these experiments.

Oblique Langmuir waves at frequencies $f = (150\text{--}350)$ MHz $< f_p$, corresponding to dispersion mode (10.52) are observed for a very long time in the plasma afterglow (up to 100–200 μs). This lifetime of the waves must be regarded as quite long since the oscillation period is not greater than several nanoseconds. Waves in the beam discharge afterglow are illustrated in figure 10.23–10.25, which show, together with the evolution of the plasma density in time, the oscillograms of the detected envelope of the HF wave amplitude.

As we see from the figures shown, the observed waves form dense bunches in which the characteristic time of variation of the oscillation amplitude envelope is several microseconds, that is, of the order of a thousand oscillation periods. Therefore, these wave bunches are, as in the earlier experiments [10.60, 10.61], solitary HF-filled waves.

It is not difficult to use the oscillograms given above to evaluate the velocity of motion of wave bunches through the plasma, as a quotient of the distance L from the HF-field and plasma-density indicators to the

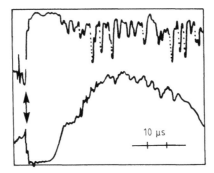

Figure 10.24 Oscillograms similar to figure 10.23, $L = 110$ cm, $v = 2 \times 10^6$ cm s^{-1}. Examples of different correlation between electric field bunches and plasma density wells [10.74, 10.75].

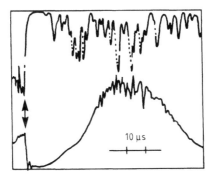

Figure 10.25 Oscillograms similar to figure 10.23, $L = 110$ cm, $v = 2 \times 10^6$ cm s^{-1}. More examples of different correlation between electric field bunches and plasma density wells [10.74, 10.75].

plasma source chamber, to the flight time which is approximately equal to the interval between the moment of turning off the electron beam and the moment of detection of the bunches and of the plasma (of a given density). The figures show that the wave bunches move together with the plasma at a velocity $v = (2–6) \times 10^6$ cm s^{-1}. The characteristic sizes of wave bunches and distances between them are found by multiplying this velocity (different in different situations) by the duration of wave pulses and intervals between them.

The waves observed are characterized by steep modulation of their amplitude and by clearly pronounced spatial periods: $\lambda_M = (5–6)$ cm $= \lambda_0/2$ and $\lambda_M = 10–12$ cm $\simeq \lambda_0$, where $\lambda_0 = u/f$ is the wavelength of the original wave generated by the beam in the plasma (at $u = 2.5 \times 10^9$ cm s^{-1} and $f = 2.5 \times 10^8$ s^{-1}, we find that $\lambda_0 = 10$ cm). The characteristic longitudinal size of wave bunches is about 4 cm, which is of the order of magnitude of one diameter of the plasma column.

The beam plasma afterglow reveals a modulation of plasma density, observed as a series of wells and bunches, which roughly correlate with the electric field wave bunches. In some conditions, wave field bunches correspond mostly to plasma density wells (see figure 10.24), while in other situations some field bunches rather correspond to plasma density bunches (see figure 10.25); on the whole, the pattern of wave field and plasma density modulation looks similar to well-developed plasma turbulence (Langmuir turbulence) [10.58, 10.64]. In view of the arguments given above and the experience of the preceding experiments [10.59, 10.60, 10.61, 10.73], we can regard the wave objects observed as oblique Langmuir solitons with HF filling.

Wave bunches suffer no spreading when they propagate along the set-up. We see this, for instance, in figures 10.26–10.29, which show oscillograms recorded at two probes located at different distances from the plasma source; the distance between the probes was 94 cm. The data of figures 10.26, 10.28 and 10.29 were obtained in a single 'shot' of the beam using two independent receivers of oscillations; the receivers were tuned to the same frequency. We see that the observed wave bunches, moving along the magnetic field, cover a distance of about 100 cm in around 15 μs; no spreading is observed; the width of individual pulses, their mutual positions and the total width of the wave packet do not change as the packet travels along the set-up. The observed wave packets are therefore essentially non-linear, since a linear wave packet spreads very rapidly, over distances of the order of its width (see section 10.2).

Wave bunches are localized in the plasma: they are nearly stationary with respect to the plasma and move together with it with respect to the set-up. The velocity of motion of wave bunches and the plasma is close to the ion-sound velocity in atomic hydrogen: $c_s = (2\text{–}5) \times 10^6$ cm s^{-1} (the upper limit corresponds to $T_e = 20$ eV). The conclusion about the localization of the observed non-linear waves in the plasma was obtained by comparing their propagation velocity relative to the set-up with the velocity of plasma propagation. Figure 10.26 compares the time shifts of the oscillograms of two HF probes spaced by a distance of 94 cm, and the time shifts of two plasma density indicators spaced by the same distance. We see that the time of flight over this distance is 15 μs both for wave bunches and for the plasma itself; the velocity of wave motion together with the plasma is (on average) $v \simeq 6 \times 10^6$ cm s^{-1}.

Results similar to those shown in figure 10.26 were also obtained at lower plasma flow velocity: $v \simeq 2 \times 10^6$ cm s^{-1} (see figure 10.27). The oscillograms of figure 10.27 were recorded using the same wave field indicator which was shifted from the initial position to different distances in the 'plasma flow direction'. In contrast to figure 10.26, these oscillograms were recorded during different beam pulses. Under the conditions of figure 10.27, wave

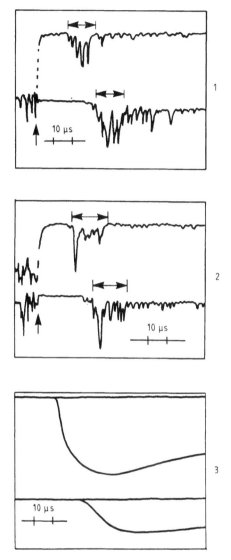

Figure 10.26 Oscillograms of signals of two identical electric-field indicators 1 and 2 and a plasma-density indicator 3 removed to a distance of 94 cm from one another; $f = 250$ MHz. Horizontal arrows in figures 10.26–10.29 mark the total duration (total length) of the wave packet [10.74, 10.75].

bunches again propagate without suffering pulse-shape spreading, and we notice that their time of flight along the set-up is almost equal to that of the plasma itself; hence, the wave bunches are, as before, 'frozen' into the plasma.

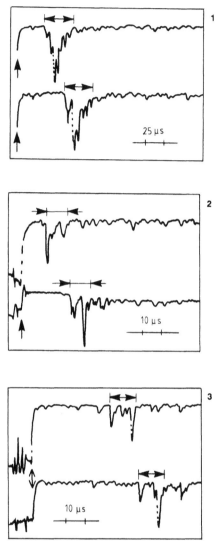

Figure 10.27   Oscillographic traces of signals of one wave electric field indicator shifted from the initial position along the 'plasma flow direction' by a distance of: 1, 30 cm; 2, 50 cm; 3, 20 cm [10.74, 10.75].

The wave bunches observed are thus either stationary relative to the plasma or move with respect to it but at a velocity which is not higher than the ion-sound velocity c_s. In the case of the experiments described here, this velocity is in fact three orders of magnitude lower than the characteristic velocity of propagation of linear waves with dispersion (10.52). This

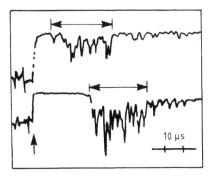

Figure 10.28   Oscillograms of signals of two wave-field indicators at a distance of 94 cm from one another [10.74, 10.75].

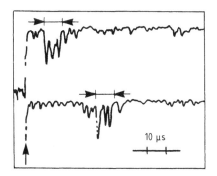

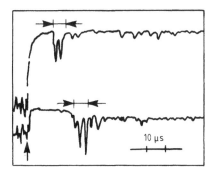

Figure 10.29   Same as in figures 10.27 and 28. Examples of self-compression of waves while they travel along the set-up. The spacing between the two probes was 80 cm [10.74, 10.75].

observation also supports the conclusion that the oblique Langmuir waves are essentially non-linear objects.

The non-spreading wave bunches observed in these experiments revealed an evolution in time which is displayed in figures 10.28 and 10.29. For instance, only 'nuclei' of peaks of electric field are seen in some places in the top oscillograms of figure 10.28, while the lower oscillograms of this figure demonstrate that 15 to 20 μs later clearly pronounced wave bunches with appreciably higher amplitude were able to grow at the same positions (shifted by 94 cm along the magnetic field). It is apparent that the lower oscillograms display a considerably larger number of clearly defined wave bunches than the upper ones. The total length of the entire set of wave bunches (indicated by arrows) is identical in the upper and lower oscillograms of figure 10.28 (and indeed in figures 10.26 and 10.29).

The self-compression of waves in the course of their propagation along the set-up (together with the plasma) is also illustrated conclusively in figures 10.26–10.29; we see that as the wave moves on, the wave bunches become considerably more pronounced; thus the width of wave bunches under the conditions of figure 10.29 is so small at the end of the plasma column that the oscilloscope ray is barely capable of resolving them. Figures 10.28 and 10.29 are therefore examples of modulational instability (self-compression) of the waves under consideration: the instability increment in the conditions of these figures is $\gamma \simeq 10^5$ s$^{-1} \simeq f_\mathrm{p} m/M$, where m and M are the electron and proton mass, respectively.

The lifetime τ of the observed packets of oblique Langmuir waves (solitons) in the free path regime (i.e. without pumping) is obviously greater than 20–30 μs, that is, greater than 10^4 oscillation periods. During this period of time, solitons travel—without spreading—nearly the entire length of the set-up.

As for the mechanism of self-compression of non-linear oblique Langmuir waves, which is responsible for the generation of the solitons observed in this study, the experimental data support the following interpretation. As has been shown above, the characteristic wave modulation length λ_M is less than, or of the order of, the initial wavelength (i.e., the HF filling 'collapses' abruptly in the process of modulation); hence, the mechanism of wave self-compression may have the same nature as in the experiments [10.59, 10.61] where it was caused by the short-wavelength variant of the modulational instability (see section 10.4). It is unlikely that this mechanism is related to the familiar Lighthill criterion (10.15) which is valid for the formation of the envelope soliton ($\lambda_\mathrm{M} \gg \lambda_0$), that is, for the situation in which the HF filling of a wave packet changes adiabatically.

The frequency spectrum of oscillations (which are synchronous at low frequencies) in the solitons observed has a bandwidth of about 40 MHz, even though the total spectrum width exceeds 200 MHz. A clear time delay is observed between the formation of oscillation bunches in various parts of the frequency spectrum: lower-frequency oscillations arrive considerably later at a remote HF probe. For instance, waves at a frequency 150 MHz arrive at the probe at about 100 cm from the plasma source 10 to 12 μs later than the 200 MHz waves, and about 20 μs later than the 250 MHz waves, and so on [10.74]. This fact invites an interpretation of the cascade pumping of waves in the plasma, as resulting from collective processes of wave scattering and decay [10.1].

The paper [10.74] was thus the first to discover slow (almost stationary with respect to the plasma) long-lived oblique Langmuir solitons which arise as a result of the modulational instability of waves (10.52). The experiments discussed above show that these solitons are stable. This property corresponds to the theory [10.89].

Structures of the type of oblique Langmuir waves were also experimentally observed in [10.90].

On non-linearities responsible for the formation of oblique Langmuir solitons, see [10.91].

The conditions of existence and stability of the Langmuir soliton were discussed in detail in [10.92].

10.6   Fast Solitons in Magnetized Plasma Waveguides

One of the features of the oblique Langmuir soliton described in section 10.5 is its low velocity: even if its velocity with respect to the plasma is non-zero, it is still much lower than the ion-sound velocity. It is shown, however, that this is not the only soliton that can exist on this branch of waves. The point is that with this type of (linear) dispersion which corresponds to waves (10.52) (and also to ion-acoustic waves and waves on 'shallow water'), there exist solitary non-linear waves (solitons) described by the (KDV) equation. These KDV-type solitons are non-spreading travelling pulses which arise as a result of the equilibrium which sets up between the non-linear steepening of the wavefront and its dispersion spreading [1.15, 10.3, 10.4]. In contrast to solitons which result from the modulational instability, these solitons are not HF-filled.

The velocity of a KDV-type soliton is naturally a function of its amplitude. By an order of magnitude, however, it is close to the phase velocity v_{ph} of a linear wave in that interval of wave numbers (k) where v_{ph} depends most strongly on k, that is, at the bend on the dispersion curve $\omega(k)$. The velocity of an ion-acoustic soliton thus is not very much higher than the ion-sound velocity, and that of a space-charge solitary wave on branch (10.52) is similarly not much higher than the characteristic 'plasma-waveguide' velocity $v_{ph} = \omega a$ where a is the plasma column radius. The longitudinal size of the soliton also corresponds to the wave number in the region of the strongest dispersion of the phase velocity. The longitudinal size of a space-charge KDV-type soliton in a magnetized plasma waveguide is thus of the order of a.

Space-charge solitons in a radially confined magnetized plasma column were experimentally observed in [10.92–10.94], and ion-acoustic solitons, in

Table 10.1   Properties of solitons in magnetized plasma waveguide.

Type of solitary wave	Propagation velocity relative to plasma (by order of magnitude)	Structure	Excitation mechanism
Oblique Langmuir (slow) soliton	$v \ll c_s$	'Langmuir -filled' density well	Modulational instability
Space-charge soliton on the (10.52) branch ('fast' soliton)	$v \simeq 10^3 c_s$	No filling	KDV mechanism

[10.38].

Table 10.1 gives a comparison of the external properties of two types of soliton observed on the wave branch (10.52): the oblique Langmuir soliton described in section 10.4 and the KDV-type soliton formed by a space-charge wave. For the velocity of the KDV-type soliton, the table gives the value $v \simeq 10^3 c_s$, which corresponds to the specific conditions chosen in [10.92]. This table demonstrates that the two types of soliton differ considerably in all their parameters.

An interesting example of soliton variety on this branch was described in a theoretical paper [10.95]. I mean a soliton on the slow space-charge wave in a non-neutralized cylindrical electron beam at a current slightly below the limiting Bursian current (2.9), (2.11). The rate of propagation of such a soliton is low in comparison with the velocity of the neutralized beam, which is quite convenient for using this soliton for acceleration of ions to high energies.

It is of interest to mention in this connection the paper [7.23], cited at the end of chapter 7, in which the collapse of electron–ion waves in a quasineutral relativistic electron beam was described. A possible relation of space-charge solitons to electron beam current limitation was discussed in [10.96].

10.7   Upper-Hybrid Solitons

These solitons exist in magnetized plasma and belong to waves whose frequencies are greater than both the electron Langmuir frequency ω_p and the electron Larmor frequency ω_H. These waves are described by the dispersion equation [3.20]:

$$\frac{\omega_p^2}{\omega^2}\frac{k_z^2}{k^2} + \frac{\omega_p^2}{(\omega^2 - \omega_H^2)}\frac{k_\perp^2}{k^2} = 1 \tag{10.53}$$

where k_z and $k_\perp$ are the longitudinal (along the field H) and the transverse wave numbers, respectively. In the approximation $\omega_p/\omega_H \to 0$, this equation describes oblique Langmuir waves illustrated in figure 10.22. There is no ion term in equation (10.53). It can be ignored if the quantity k_z^2/k^2 is not too small (as I assume hereafter). Two limiting cases are of special interest.

1. If $\omega_H^2 \gg \omega_p^2$, then

$$\left.\begin{aligned} \omega &= \omega_p \cos\Theta = \frac{\omega_p k_z}{(k_z^2 + k_\perp^2)^{1/2}} \\ \omega &= \omega_H + \frac{\omega_p^2}{2\omega_H}\sin^2\Theta \end{aligned}\right\}. \tag{10.54}$$

This oscillation branch is known, together with equation (10.52), as the *Trivelpiece–Gould mode* [10.97].

2. If $\omega_H^2 \ll \omega_p^2$, then

$$\left.\begin{aligned} \omega &= \omega_H \cos\Theta \\ \omega &= \omega_p + \frac{\omega_H^2}{2\omega_p}\sin^2\Theta \end{aligned}\right\}. \tag{10.55}$$

The qualitative features of these oscillation branches are also retained when the frequencies ω_p and ω_H are not very different. These oscillation branches are shown in figure 10.22 where $k_\perp$ is approximated by assuming $k_\perp \propto 1/a$, and a is the radius of the plasma column.

Let us turn now to figure 10.22(a) which represents magnetized electron oscillations that we were treating so far. If $k_z a \gg 1$, that is, if, for instance, $\lambda_z = 2\pi/k_z \ll a$, we find that $\omega \simeq \omega_p$ on the lower oscillation branch. In fact, this was to be expected: oscillations whose wavelength is less than the plasma-column radius occur at the Langmuir frequency. We have seen above that when Langmuir oscillations are excited by a low-density electron beam, the resonance condition $\omega_p \simeq ku$ is satisfied. In other words, the working point in the beam excitation of oscillations is determined by the intersection of the branch in figure 10.22(a) with the straight line $\omega \simeq ku$. If, however, $k_z a \ll 1$, that is, if $\lambda_z \gg a$, then we have on the lower branch

$$\frac{\omega}{k_z} = \frac{\omega_p}{k} \simeq \frac{\omega_p}{k_\perp} \simeq \omega_p a. \tag{10.56}$$

Therefore, the fixed quantity in the oscillations in which the wavelength is much greater than the plasma column radius, is no longer the Langmuir frequency but the phase velocity ω/k_z; as for the frequency, it is proportional to the wave number k_z. This result has a clear physical meaning: if $\lambda_z \gg a$, then the 'elasticity' of the plasma with respect to the longitudinal displacement of magnetized electrons (which determines the oscillation frequency) is related not to the total electric field $\boldsymbol{E}$ of the separated charges but only to its longitudinal component $\boldsymbol{E}_z = \boldsymbol{E}k_z/k$, or rather to the derivative of E_z along z; as a result, it is found to be proportional to the factor k_z^2/k^2. It then follows that $\omega^2 = \omega_p^2 k_z^2/k^2$, whence $\omega/k_z = \omega_p/k = \omega_p/a = $ constant. This immediately implies that if the beam velocity $u > \omega_p/k \simeq \omega_p a$, the beam cannot excite electron oscillations of the plasma on the lower branch in figure 10.22(a): the line $\omega = k_z u$ does not intersect this branch. The upper branch in figure 10.22(a) gives the frequency band from the Larmor frequency ω_H to the upper hybrid frequency $\omega = (\omega_H^2 + \omega_p^2)^{1/2}$; if $ak_z \gg 1$, the latter frequency coincides with ω_H. Likewise, the upper branch of figure 10.22(b) demonstrates the frequency band: from the Langmuir frequency ω_p to the upper hybrid frequency $\omega = (\omega_H^2 + \omega_p^2)^{1/2}$; if $ak_z \gg 1$, the latter frequency coincides with ω_p. Another interesting property of long-wavelength oscillations ($\lambda_z \gg a$, $k_z a \ll 1$) is that their group velocity in the z direction is approximately equal to the phase velocity:

$$v_g \simeq \frac{\omega_p}{k} \simeq v_{\text{ph}} = \frac{\omega}{k_z} \simeq u. \tag{10.57}$$

This result follows only from the transverse boundedness of the system: the group velocity of (Langmuir) oscillations in unbounded plasma does not vanish only because of the thermal motion of plasma electrons, and is very small in comparison with the beam propagation velocity. The law of dispersion of Langmuir waves, (10.1), yields

$$v_{gz} \equiv \frac{\partial \omega}{\partial k_z} = v_e^2 k_z/\omega \tag{10.58}$$

whence

$$v_{\text{ph}z}\, v_{gz} = v_e^2 = 3T_e/m \tag{10.59}$$

and

$$v_{gz} = v_e^2/u \tag{10.60}$$

where v_e is the thermal velocity of plasma electrons. We see that if $v_e \ll u$, the group velocity of Langmuir waves in unbounded plasma (v_{gz}) is much lower than the thermal velocity of plasma electrons. In contrast to this case, in bounded plasma (at $\lambda_z \gg a$), we find

$$v_{gz} \simeq u \gg v_e.$$

According to the theory [10.89], electron solitons are also possible on the branch of upper-hybrid waves. According to the data of [10.98], they are indeed observed in magnetized plasma. The experiment [10.98] was carried out in previously created plasma which was subjected to irradiation by high-power electromagnetic wave whose frequency was close to the upper-hybrid plasma frequency.

The parameters of the system: plasma column diameter about 4 cm, column length about 125 cm, magnetic field strength $B_0 \lesssim 3.5$ kG, plasma electron temperature $T_e = 2.5$ eV, plasma density $n_e \simeq 2 \times 10^{12}$ cm^{-3}, working gas (argon) pressure $p = (4–10) \times 10^{-4}$ mm Hg. Under typical conditions, the Langmuir frequency ω_p is several times greater than the Larmor frequency, which is in agreement with figure 10.22(b). The pumping wave frequency was $\omega_0/2\pi = 9.4$ GHz, the pumping power reached 50 kW and the pulse duration was from 0.3 to 1.5 μs.

A series of experiments [10.98] demonstrated that if the pumping power P does not exceed a threshold P_{cr} (about 5 kW), the radial plasma density profile remains the same as in the absence of the pumping wave (figure 10.30(a)). If, however, $P \gg P_{cr}$, a density 'well' of radial width from 3 to 4 mm (about $300r_D$) and about 25 mm long (in the direction of the column axis) appears on the density profile; the well length is almost equal to the size of the waveguide through which the pumping wave was fed. Experiments showed that the amplitude $\delta n/n$ of the well rose as the pumping power was increased and reached $\delta n/n_0 \simeq 0.25$ at $P \simeq 30$ kW. The electric field of the pumping wave concentrates in the density well (figure 10.30(b)). The frequency of this field corresponds to the upper-hybrid plasma frequency. The authors of [10.98] are of the opinion that they observed an upper-hybrid soliton (or caviton).

The following result, connected with the high frequency of electron collisions is typical both for this experiment and for those described in section 10.3: after the pumping field was turned off, the electric field captured by the density well damped out rapidly (even though the lifetime of the density well itself was obviously considerably longer than that of the electric field). We conclude that in this case again the soliton was not observed in its *free state* (after the pumping field is turned off). As a result, the propagation velocity of the soliton was unknown. This object should rather be referred to as a caviton or spikon. (On 'true' solitons observed in the free flight regime, see sections 10.4 and 10.5.)

The effect of self-compression in two frequency bands—the Trivelpiece–Gould branch (10.52) and the upper hybrid branch (10.53)—in the electron beam–plasma system was also reported in [10.99].

The interpretation of the radial collapse of upper-hybrid electron waves and of the formation of wave 'filaments', which stretch along the magnetic field, is given in [10.100].

An interesting version of the Langmuir soliton—the so-called 'auto-

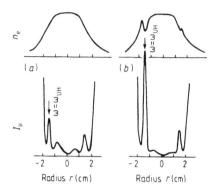

Figure 10.30   Radial profiles of wave intensity I_μ and plasma electron density. (*a*) Low-power incident wave: $P_\mu = 0.5$ kW; (*b*) High-power incident wave: $P_\mu = 30$ kW (pulse duration $\tau_\mu = 350$ ns). The pumping wave frequency is equal to the upper hybrid frequency. The quantity I_μ is measured 200 ns after turning on the pumping pulse; n_e is measured 150 ns after the end of the pulse. $B_0 = 0.6$ kG, the initial plasma density $n_{e0} = 1.5 \times 10^{12}$ cm^{-3} and the pressure $p = 5 \times 10^{-4}$ mm Hg [10.98].

oscillation soliton'—is described in [10.101].

10.8   On Ionization-Wave Solitons in Gas Discharge

It is well known (see, e.g., a recent monograph [10.102]) that ionization waves, known as discharge striations, exist in glow discharge. Theoretical papers [10.103–10.105] showed that under certain conditions, ionization waves may undergo modulational instability and even produce solitons. The modulational instability of this type is described by the same 'non-linear Schrödinger equation' which also implies the existence of the Langmuir soliton on a very different wave branch. The modulational instability may result in the formation of an ionization-wave soliton. Such solitons were observed in [10.105]. They had very specific size and oscillation frequencies that are typical for ionization waves.

An interesting observation [10.106] proved the existence of a strongly supersonic soliton, namely, a plasma density hump, of approximately Gaussian shape, which propagates towards the anode at a velocity almost equal to the thermal velocity of plasma electrons.

10.9   On Langmuir Solitons in Outer Space

The Voyager-1 and 2 spacecrafts on the flight to Jupiter in 1979 recorded intriguing data which pointed to the possibility of generation of Langmuir

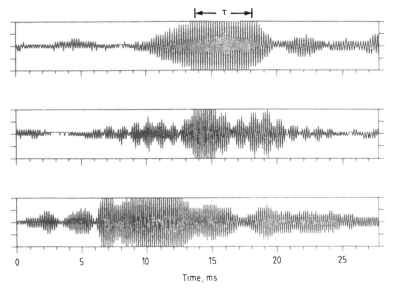

Figure 10.31 Oscillograms of Langmuir wave electric field [10.107]. $\tau \rightarrow 100r_D$.

solitons in the solar wind. Figure 10.31 shows an oscillogram of the electric field of Langmuir waves (at the electron plasma frequency $f = f_p = 5.6$ kHz) excited in the solar wind plasma by an electron beam of about 25 keV energy; the beam is injected into this plasma from the diurnal side of the Jovian ionosphere (plasma density 0.5 cm^{-3}). According to the authors' interpretation [10.107], this result signifies the formation of a cavity which captured bunched electric field of Langmuir waves; the cavity is stationary with respect to the plasma and propagates with it at the velocity of the solar wind (about $4 \times 10^7 \text{ cm s}^{-1}$). In accordance with the discussion in the preceding paragraph, the time scale of modulation of the electric field can be converted into the spatial scale, whereby we learn that the cavity width (about 50 ms) is about $100r_D$, where the Debye radius $r_D = 20$ m (the electron temperature $T_e = 5$ eV). The results of measurements of electric fields in plasma waves and of the plasma parameters fit satisfactorily the relation

$$E^2/16\pi nT \simeq (\kappa r_D)^2$$

where $E \approx 10^{-4} \text{ V cm}^{-1}$ and $\kappa = \pi/100r_D$.

The authors of [10.7] and [10.107] are of the opinion that these measurements gave the first evidence of the formation of Langmuir envelope solitons (i.e., the evidence of modulational instability) in the solar wind.

The very intriguing fact that a beam of electrons, at energies of tens of keV, pierces the solar wind plasma and travels without appreciable energy losses at least one half of the distance from the Sun to the Earth, is thus explained in terms of the modulational instability, which reduces the phase velocities of Langmuir waves and knocks them out of Čerenkov resonance with the electron beam (see section 10.2).

This putative interpretation of non-linear Langmuir waves in extraterrestrial plasma in terms of solitons is also supported by the author of the review paper [1.21] on strong Langmuir turbulence.

As for strong Langmuir turbulence itself, it has been analysed in a very large number of papers. Among the most recent ones I can refer the reader to the papers [10.108–10.114], which treat both theory and experiment.

To conclude this chapter, I will point out a very unorthodox method, developed in [10.115], of experimental analysis of the modulational instability of Langmuir waves in the plasma. It is connected with the so-called monotrone mechanism of generation of HF oscillations by an electron beam whose time of flight across the accelerating gap is comparable with the oscillation period (in the case in question here, this is the period of the Langmuir plasma oscillations). The authors indicated the conditions under which this mechanism may dominate the generation of oscillations caused by the 'ordinary' beam instability.

References

[10.1]   Tsitovich V N 1967 *Nonlinear Effects in Plasma* (Moscow: Nauka) chapter 4 (in Russian)

[10.2]   Vedenov A A and Rudakov L I 1964 *Dokl. AN SSSR* **159** 767 (Engl. Transl. 1964 *Sov. Phys.–Doklady* **9** 1073)

[10.3]   Kadomtsev B B and Karpman V I 1971 *Uspekhi Fiz. Nauk* **103** 193 (Engl. Transl. 1971 *Sov. Phys.–USPEKHI* **14** 40)

[10.4]   Karpman V I 1973 *Nonlinear Waves in Dispresive Media* (Moscow: Nauka) (in Russian)

[10.5]   Chen F F 1984 *Introduction to Plasma Physics and Controlled Fusion* vol 1 Plasma Physics (New York: Plenum)

[10.6]   Galeev A A, Sagdeev R Z, Shapiro V D and Shevchenko V I 1977 *Zh. Eksp. Theor. Fiz.* **72** 507 (Engl. Transl. 1977 *Sov. Phys.–JETP* **45** 266)

[10.7]   Gurnett D A and Anderson R R 1977 *J. Geophys. Res.* **82** 632

[10.8]   Goldman M V 1984 *Rev. Mod. Phys.* **56** 709

[10.9]   Vedenov A A, Velikhov E P and Sagdeev R Z 1961 *Nucl. Fusion* **1** 82

[10.10] Sagdeev R Z 1964 Collective Processes and Shock Waves in the Plasma *Voprosy Teorii Plasmy (Aspects of Plasma Theory* ed M A Leontovich No 4 (Moscow: Atomizdat) p. 20

[10.11] Tran M Q 1979 *Phys. Scr.* **20** 317

[10.12]   Mikhailovskii A B, Petviashvili V I and Fridman A M 1976 *Pis'ma v Zh. Eksp. Teor. Fiz.* **24** 53 (Engl. Transl. 1976 *JETP-Lett.* **24** 43)

[10.13]   Lonngren K 1983 *Plasma Phys.* **25** 943

[10.14]   Nakamura Y 1972 *IEEE Trans.* **PS-10** 180

[10.15]   Hasegawa A 1985 *Adv.Phys.* **34** 1

[10.16]   Mikhailovskii A B 1986 *Nonlinear Phenomena in Plasma and Hydrodynamics* ed R Z Sagdeev (Moscow: Mir)

[10.17]   Antipov S V, Nezlin M V, Rodionov V K, Rylov A Yu, Snezhkin E N, Trubnikov A S and Khutoretsky A V 1988 *Fizika Plasmy* **14** 1104 (Engl. Transl. 1988 *Sov. J. Plasma Phys.* **14** 648)

[10.18]   Nozaki K 1981 *Phys. Rev. Lett.* **46** 184

[10.19]   Makino M, Kamimura T and Taniuti T 1981 *J. Phys. Soc. Japan* **50** 980

[10.20]   Ivanov A A 1975 *Fizika Plasmy* **1** 147 (Engl. Transl. 1975 *Sov. J. Plasma Phys.* **1** 78)

[10.21]   Ivanov A A and Leiman V G 1978 Electron Beams in Plasmachemistry *Khimiya Plasmy (Chemistry of the Plasma)* ed B M Smirnov (Moscow: Atomizdat) (in Russian)

[10.22]   Rudakov L I and Tsytovich V N 1978 *Phys. Repts.* **40** 1

[10.23]   Degtyarev L M, Sagdeev R Z, Soloviev G I, Shapiro V D and Shevchenko V I 1989 *Zh. Eksp. Theor. Fiz.* **95** 1690 (Engl. Transl. 1989 *Sov. Phys.-JETP* **68** 975)

[10.24]   Doolen G D, DuBois D F and Rose H A 1986 *Phys. Rev. Lett.* **54** 804

[10.25]   Dyachenko A I, Zakharov V E, Rubenchik A M, Sagdeev R Z and Shvets V F 1988 *Zh. Eksp. Theor. Fiz.* **94** 144 (Engl. Transl. 1988 *Sov. Phys.-JETP* **67** 513)

[10.26]   Scott A 1970 *Active and Nonlinear Wave Propagation in Electronics* (New York: Wiley, Interscience)

[10.27]   1978 *Solitons in Action* ed K Lonngren and A Scott (New York: Academic Press)

[10.28]   1978 *Solitons in Condenced Matter Physics (Symposium Proc.)* ed A R Bishop (New York: Springer)

[10.29]   Rebbi C 1979 Solitons *Sci. Amer.* **240** 76

[10.30]   Lamb G L Jr 1980 *Elements of Soliton Theory* (New York: Wiley)

[10.31]   1980 *Solitons* ed R K Bullough and P J Caudrey (Heidelberg: Springer)

[10.32]   Zakharov V E, Manakov S V, Novikov S P and Pitaevsky L P 1980 *Soliton Theory* (Moscow: Nauka)

[10.33]   Newell A C 1985 *Solitons in Mathematics and Physics* (Society for Industrial and Applied Mathematics, University of Arizona)

[10.34]   Dodd R K, Eilbeck J C, Gibbon J D and Morris H C 1984 *Solitons and Non-Linear Wave Equations* (London:Academic)

[10.35] Hefter E F and Mitropolsky I A 1991 *J. Moscow. Phys. Soc.* **1** 99

[10.36] Davydov A S 1984 *Solitons in Molecular Systems* (Kiev: Naukova dumka) (in Russian)

[10.37] Sagdeev R Z 1958 *Plasma Physics and the Problem of Controlled Thermonuclear Fusion Reactions* ed M A Leontovich **4** (Moscow: AN SSSR) p. 384

[10.38] Ikezi H 1970 *Phys. Rev. Lett.* **25** 11

[10.39] Lighthill M S 1965 *T. Inst. Math. Appl.* **1** 269

[10.40] Benjamin T B and Feir J E 1967 *J. Fluid Mech.* **27** 417

[10.41] Hammack J L and Segur H 1974 *J. Fluid Mech.* **65** 289

[10.42] Chiao R S, Gardmire F and Townes C H 1964 *Phys. Rev. Lett.* **13** 479

[10.43] Askaryan G A 1973 *Uspekhi Fiz. Nauk* **III** 249 (Engl. Transl. 1974 *Sov. Phys.-USPEKHI* **16** 680)

[10.44] Bespalov V I, Litvak A G and Talanov V I 1968 *Nonlinear Optics* ed R V Khokhlov (Novosibirsk: Nauka) p. 428

[10.45] Kuznetsov E A, Rubenchik A M and Zakharov V E 1986 *Phys. Repts.* **142** 103

[10.46] Rasmussen J J and Rypdal K 1986 *Phys. Scr.* **33** 481, 498

[10.47] Zakharov V E 1972 *Zh. Eksp. Theor. Fiz.* **62** 1745 (Engl. Transl. 1972 *Sov. Phys.-JETP* **35** 908)

[10.48] Rudakov L I 1972 *Dokl. AN SSSR* **207** 821 (Engl. Transl. 1973 *Sov. Phys.-Doklady* **17** 1166)

[10.49] Nezlin M V 1986 *Uspekhi Fiz. Nauk* **150** 3; 1986 *Priroda* **10** 23 (Engl. Transl. 1986 *Sov. Phys.-USPEKHI* **29** 807)

[10.50] Nezlin M V and Snezhkin E N 1992 *Rossby Vortices and Spiral Structures* (Heidelberg, New York: Springer–Verlag)

[10.51] Gaponov A V and Miller M A 1958 *Zh. Eksp. Theor. Fiz.* **34** 242 (Engl. Transl. 1958 *Sov. Phys.-JETP* **7** 168)

[10.52] Trubnikov B A 1972 *Zh. Eksp. Theor. Fiz.* **62** 971 (Engl. Transl. 1972 *Sov. Phys.-JETP* **35** 513)

[10.53] Zakharov V E and Rubenchik A M 1972 *Prikl. Mekh. i teor. fiz.* **5** 84

[10.54] Galeev A A and Sagdeev R Z 1973 Nonlinear Plasma Theory *Voprosy Teorii Plasmy (Problems of Plasma Theory)* no 7 ed M A Leontovich (Moscow: Atomizdat)

[10.55] Nezlin M V 1981 *Proc. XY Intern. Conf. Phys. Ionized Gases* (Minsk) Invited papers p. 279

[10.56] Nishikava K 1968 *J. Phys. Soc. Japan* **24** 916, 1152

[10.57] Silin V P 1973 *Parametric Effects of High-Power Irradiation on the Plasma* (Moscow: Nauka)

[10.58] Andreev N A, Kirii A Yu and Silin V P 1969 *Zh. Eksp. Theor. Fiz.* **57** 1024 (Engl. Transl. 1970 *Sov. Phys.-JETP* **30** 559)

[10.59]   Wong A Y and Quon B H 1975 *Phys. Rev. Lett.* **34** 1499

[10.60]   Antipov S V, Nezlin M V, Snezhkin E N and Trubnikov A S 1978 *Zh. Eksp. Theor. Fiz.* **74** 965 (Engl. Transl. 1978 *Sov. Phys.-JETP* **47** 506)

[10.61]   Antipov S V, Nezlin M V, Snezhkin E N and Trubnikov A S 1979 *Zh. Eksp. Theor. Fiz.* **76** 1571 (Engl. Transl. 1979 *Sov. Phys.-JETP* **49** 797)

[10.62]   Rypdal K and Rasmussen J J 1989 *Phys. Scr.* **40** 192

[10.63]   Karfidov D M, Rubenchik A M, Sergeichev K F and Sychev I A 1988 *Pis'ma v Zh. Eksp. Teor. Fiz.* **48** 315 (Engl. Transl. 1988 *JETP-Lett.* **48** 346); 1990 *Zh. Eksp. Theor. Fiz.* **98** 1592 (Engl. Transl. 1990 *Sov. Phys.-JETP* **71** 892)

[10.64]   Morales G I and Lee Y C 1976 *Phys. Fluids* **19** 690

[10.65]   Mofiz U A, DeAngelis U, Shukla P K and Rao N N 1985 *Phys. Fluids* **28** 826

[10.66]   Degtyarev L M, Makhankov V G and Rudakov L I 1974 *Zh. Eksp. Theor. Fiz.* **67** 533 (Engl. Transl. 1974 *Sov. Phys.-JETP* **40** 264)

[10.67]   Ikezi H, Nishikawa K and Mima K 1974 *J. Phys. Soc. Japan* **37** 766

[10.68]   Kato K, Kunishima A and Minami K 1989 *Japan. J. Appl. Phys.* **28** 512

[10.69]   Cheung P Y, Wong A Y, Tanikawa T, Santoru J, DuBois D F, Rose A H and Russel D 1989 *Phys. Rev. Lett.* **62** 2676

[10.70]   Larroshe O and Pesme D 1990 *Phys. Fluids* **B2** 1751

[10.71]   Bauer B S , Wong A Y, Scurry L and Decyk V K 1990 *Phys. Fluids* **B2** 1941

[10.72]   Kato K and Kunishima A 1989 *Japan. J. Appl. Phys.* **28** 512

[10.73]   Ikezi H, Chang R P H and Stern R A 1976 *Phys. Rev. Lett.* **36** 1047

[10.74]   Antipov S V, Nezlin M V and Trubnikov A S 1980 *Zh. Eksp. Theor. Fiz.* **78** 1743 (Engl. Transl. 1980 *Sov. Phys.-JETP* **51** 874)

[10.75]   Antipov S V, Nezlin M V and Trubnikov A S 1981 *Physica* **D3** 311

[10.76]   Wong A Y 1979 *Comm. Plasma Phys. Contr. Fus.* **5** 79

[10.77]   Cheunq P Y, Wong A Y, Darrow C B and Qian S J 1982 *Phys. Rev. Lett.* **48** 1348

[10.78]   Cheung P Y and Wong A Y 1985 *Phys. Fluids* **28** 1538

[10.79]   Cheung P Y and Wong A Y 1985 *Phys. Rev. Lett.* **55** 1880

[10.80]   Zakharov V E, Pushkarev A N, Shvets V F and Yankov V V 1988 *Pis'ma v Zh. Eksp. Teor. Fiz.* **48** 79 (Engl. Transl. 1988 *JETP-Lett.* **48** 83)

[10.81]   Astrelin V T, Breizman B N, Sedlacek Z and Jungwirth K 1988 *Fizika Plasmy* **14** 706 (Engl. Transl. 1988 *Sov. J. Plasma Phys.* **14** 417)

[10.82] Kingsep A S 1983 *Itogi Nauki i Tekhniki (Science and Technology Summary). Plasma Physics* ed V D Shafranov **4** (Moscow: VINITI) p. 48

[10.83] Robinson P A 1991 *Phys. Fluids* **3B** 545

[10.84] Astrakhantsev N V, Volkov O L, Karavaev Yu S and Kichigin G N 1985 *Phys. Lett.* **110A** 129

[10.85] McDaniel E W 1964 *Collision Phenomena in Ionized Gases* (New York: John Wiley)

[10.86] McDaniel E W 1964 *Collision Phenomena in Ionized Gases* (New York: John Wiley) Supplement p. 2

[10.87] Buchelnikova N S and Matochkin E P 1980 *Comm. Plasma Phys. and Contr. Fusion* **6** 21

[10.88] Khodataev K V and Tsytovich V N 1978 *Fizika Plasmy* **4** 799 (Engl. Transl. 1978 *Sov. J. Plasma Phys.* **4** 449)

[10.89] Kolchin A A and Kuznetsov E A 1987 *Fizika Plasmy* **13** 1204 (Engl. Transl. 1987 *Sov. J. Plasma Phys.* **13** 694)

[10.90] Bond J W 1982 *Plasma Phys.* **24** 1495

[10.91] Pečseli H L and Rasmussen J J 1980 *Plasma Phys.* **22** 421

[10.92] Lynov J P, Michelsen P, Pečseli H L et al. 1979 *Phys. Scr.* **20** 317

[10.93] Ikezi H, Barret P J, White R B and Wong A Y 1971 *Phys. Fluids* **14** 1997

[10.94] Krivoruchko S M, Fainberg Ya B, Shapiro V D and Shevchenko V I 1974 *Zh. Eksp. Theor. Fiz.* **67** 2092 (Engl. Transl. 1974 *Sov. Phys.-JETP* **40** 1039)

[10.95] Hughes T P and Qtt E 1980 *Phys. Fluids* **23** 2265

[10.96] Rutkevich B N and Rutkevich S B 1990 *Fizika Plasmy* **16** 683 (Engl. Transl. 1990 *Sov. J. Plasma Phys.* **16** 396)

[10.97] Krall N A and Trivelpice A W 1973 *Principles of Plasma Physics* New York: McGraw Hill)

[10.98] Cho T, Yamazaki K and Tanaka S 1982 *J. Phys. Soc. Japan* **51** 988

[10.99] Chistiansen P J, Jain V K and Stenflo L 1981 *Phys. Rev. Lett.* **46** 1333

[10.100] Giles M J 1981 *Phys. Rev. Lett.* **47** 1606

[10.101] Kolchugina I A, Litvak A G and Sergeev A M 1982 *Pis'ma v Zh. Eksp. Teor. Fiz.* **35** 510 (Engl. Transl. 1982 *JETP-Lett.* **35** 631)

[10.102] Raiser Yu P 1991 *The Physics of Gas Discharge* (Heidelberg: Springer)

[10.103] Bekki N 1981 *J. Phys. Soc. Japan* **50** 659

[10.104] Lagarkov A N and Rutkevich I M 1981 *Fizika Plasmy* **7** 1132 (Engl. Transl. 1981 *Sov. J. Plasma Phys.* **7** 622); 1979 *Dokl. AN SSSR* **249** 593 (Engl. Transl. 1979 *Sov. Phys.-Doklady* **24** 933); 1983 *Teplofiz. Vysok. Temp.* **21** 433 (Engl. Transl. 1983 *High Temp.* **21** 321)

[10.105] Ohe K and Hashimoto M 1984 *Phys. Fluids* **27** 1863

[10.106] Sa A B and Mendosa J T 1984 *Intern. Conf. on Plasma Phys.* **2** (Lausanne:) p. 284

[10.107] Gurnett D A, Maggs J E, Gallagher D L, Kurth W S and Scarf F L 1981 *J. Geophys. Res.* **86A** 8833

[10.108] Robinson P A and Newman D L 1990 *Phys. Fluids* **B2** 2999, 3017, 3120; 1989 *Phys. Fluids* **B1** 2319

[10.109] Newman D L, Winglee R M, Robinson P A, Glanz J and Goldman M V 1990 *Phys. Fluids* **B2** 2600

[10.110] Goldstein P P and Roznus W 1990 *Phys. Fluids* **B2** 44

[10.111] Main W and Benford G 1989 *Phys. Fluids* **B1** 2479

[10.112] Benford G, Zhai X and Levron D 1991 *Phys. Fluids* **B3** 560

[10.113] Bychenkov V Yu, Karfidov D M, Sergeichev K F and Sychev I A 1991 *Pis'ma v Zh. Eksp. Teor. Fiz.* **53** 191 (Engl. Transl. 1991 *JETP-Lett.* **53** 203)

[10.114] Briand J, Berge L, Gomes A, Quemener Y, Arnas C, Armengaud M and Dinguirard J P 1990 *Phys. Fluids* **B2** 160

[10.115] Brodskii Yu Ya, Litvak A G, Slutsker Ya Z and Fraiman G M *Zh. Eksp. Theor. Fiz.* at press

11 Novel Methods of Generation and Amplification of Electromagnetic Waves by Relativistic Electron Beams

Today's highest-power generators of electromagnetic waves involve the use of high-intensity (pulsed) relativistic electron beams. One of the principles of these generators makes use of the phenomenon of formation and subsequent high-frequency oscillations of the virtual cathode in the electron beam at the super-Bursian current. The corresponding devices are known as *vircators* (and sometimes as *turbotrons*). Another principle uses the effect of relativistic transformation of the frequency of an electromagnetic wave when it is scattered on the electron beam (in other words, it uses the relativistic Doppler effect occurring in this scattering). Depending on the concentration and energy of beam electrons and on the oscillation wavelength, the scattering occurs either on individual particles ('free' electrons) or on electron waves (Raman scattering). Both types of devices are usually referred to as *free electron lasers* (or *masers*).

This chapter will describe the principles of functioning of these generators (and amplifiers) whose physics is closely related to the phenomena discussed in chapters 4 and 6.

11.1 An Electron Beam at a Current above Limiting Bursian Current as a Vircator (Turbotron)

High-frequency oscillations in two clearly defined frequency ranges are observed in the electron beam in which a virtual cathode is formed at a current above the limiting Bursian current (section 2.1). One of these two ranges is connected with the oscillations of the virtual cathode as a whole [11.1]. These oscillations [11.2–11.15] occur at 'nearly-Langmuir' frequencies

$$\omega_1 \leqslant \omega \leqslant \sqrt{2\pi}\omega_1 \tag{11.1}$$

where

$$\omega_1 = (4\pi n_1 e^2/m\gamma_0)^{1/2}$$

is the natural (Langmuir) frequency of beam electrons and γ_0 is relativistic factor (2.29, 3.47). These are the highest-frequency and highest-intensity oscillations.

Another range, at frequencies which are typically several times lower, involves the time-of-flight oscillations of beam electrons between the cathode and the virtual cathode. Owing to the physical mechanism driving these oscillations, they are analogous, in principle, to the familiar low-power Barkhauzen–Kurz oscillations which are observed in triodes with a retarding field between the positive-potential grid and the anode [11.16]. These (less intense) oscillations produce a considerable velocity spread in the electron beam and thus constitute an obstacle for the excitation of the (main) oscillations of the virtual cathode. There is, therefore, a tendency to suppress time-of-flight oscillations, using an efficient technique described below [11.4].

A plane hollow anode with a ring-shaped orifice for the electron beam is placed in front of a cylindrical hollow cathode. A hollow electron beam, forming a virtual cathode behind the anode, is formed using a longitudinal magnetic field of several tens of kOe (figure 11.1). Electrons reflected from the virtual cathode are then driven by the space-charge forces to shift perpendicularly to the magnetic field and be deposited on the anode. The loss of these electrons disrupts the generation of oscillations at time-of-flight frequencies [11.4].

Beams with electron energies of about 1 or several MeV and currents of several tens of kA (which is above the Bursian current (2.30)) in a longitudinal magnetic field of several tens of kOe thus make it possible to generate extremely high-power oscillations in pulses tens of nanoseconds long, at wavelengths from 2 to about 8 mm, with power of the order of 1–2 GW and efficiency up to several per cent [11.5–11.11]. It proved to be possible to raise the power output to 4–5 GW at somewhat higher wavelengths (4 to 10 cm) [11.13].

According to the theory (see, e.g., [11.6]), the efficiency of such generators can be increased to 30–40%, at the power output level of 10 GW and at wavelength of several centimeters.

The field strength of oscillations in the electron beam of a vircator may reach several MV cm^{-1}.

As an example, I will list the parameters of a high-power generator known as Reditron (reflect electron discriminator tube) [11.13, 11.14]—see figure 11.1. The name alludes to the removal of electrons reflected from the virtual cathode.

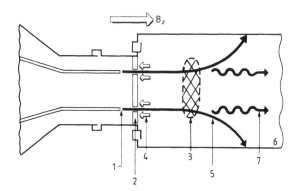

Figure 11.1 Schematic diagram of an HF generator (vircator) on a high-power relativistic electron beam with virtual cathode [11.13]. 1, hollow cathode; 2, thick anode with ring-shaped orifice, 3, virtual cathode; 4 and 5, beam electrons reflected away from the virtual cathode and those transmitted through it, respectively; 6, waveguide; 7, radiation.

The generator has a hollow cathode $2a = 6$ cm in diameter and a plane anode with a ring-shaped orifice opposite the cathode, at a distance which can be varied from 1 to 3.7 cm. The beam propagates in a tube of diameter $2R = 18$ cm, in a longitudinal magnetic field of around 5 kG; the residual gas pressure is $p \simeq 3 \times 10^{-5}$ mm Hg. The electron beam energy: $W_1 = 1.5$–2 MeV; the beam current: $I = 10$–50 kA; the radiation pulse length: 50 ns.

The generator operates at a current

$$I > I_{\mathrm{B}} = 8.5 \frac{(\gamma^{2/3} - 1)^{3/2}}{\ln(R/a)} \text{ kA}$$

at which a virtual cathode forms in the beam (behind the anode—see figure 11.1). The oscillations of the virtual cathode occur at frequencies (11.1) which can be varied in the interval from 2 to 40 GHz; the radiation emission power at 8 mm wavelength is 1.4 GW. The efficiency of the device is several per cent; according to numerical simulations [11.13], it can be increased to 40%, and the longitudinal electric field in the virtual cathode may reach around 7 MV cm^{-1}. The discrimination of reflected electrons enhances the output power of the generator by an order of magnitude. Along with the 'free' electron maser (FEM) (rather, the collective Raman scattering maser, see next section), the reditron is the highest-power electromagnetic wave generator in the centimetre waveband.

The highest energy output (up to 400 J at 30 cm wavelength in a 150 ns long pulse) was recorded at the record-power 'Aurora' electron accelerator [11.17].

The highest power output (up to 7.5 GW in a 10 ns-long pulse, at 25 cm wavelength) was also reported in [11.17].

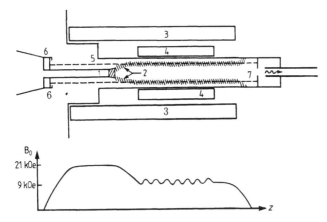

Figure 11.2   Schematic diagram of FEL (more precisely, free electron maser) using an undulator: 1, cathode; 2, electron beam; 3, solenoid; 4, undulator; 5, ray trajectory; 6, 7, mirrors.

11.2   Compton Scattering of Photons by Relativistic Electrons (Theory)

An approach based on using electron beams (mostly relativistic beams) for high-efficiency generation (and amplification) of electromagnetic oscillations in a prescribed, easily controlled range of wavelengths has taken shape in recent years and showed impressive progress; the output wavelength may vary from centimetres to x-ray wavelengths [11.18–11.21]. The devices based on a combination of Thomson (Compton) scattering of electromagnetic waves (photons) by electrons with the relativistic Doppler effect are known as *free electron masers or lasers*; the customary abbreviations are FEM (for the millimetre and submillimetre wavelength ranges) and FEL (for shorter wavelengths, from infrared to x-ray). Quite often, the term FEL is used for generators (amplifiers) of both types. Obviously, the combination 'free electron laser (maser)' is a rather crude exaggeration: first, even though beam electrons in such devices are not bound in atoms, they are not free either but move in electromagnetic fields; second, electromagnetic waves are scattered not only by individual electrons but also, in the general case, by electron waves. Furthermore, there is a whole class of especially high-power FEL in which the main effect of scattering occurs as a result of active interaction between waves with both signs of energy; the principal element of this interaction is negative-energy waves, namely, slow space-charge waves or slow cyclotron waves. Such devices are very suitable topics for this review since their principles make use of the instabilities of negative-energy waves treated in chapter 4.

Such devices are often referred to as Raman, or magneto-Raman, FEL†. They are realized in the hydrodynamic regime of the beam instability, when the velocity spread of beam particles is sufficiently small, that is, satisfies condition (4.37). In this regime, all beam electrons participate in the scattering *coherently* and the scattering intensity is greatly enhanced. Therefore, this regime of FEL is especially efficient. Condition (4.37) can be met if the beam density is sufficiently high and the wavelength is not too small (see also (1.3)).

It is possible to write simple qualitative criteria which allow one to determine in what regime a specific FEL operates. For example, the regime of scattering by individual electrons will be dominant if the time of flight of a beam electron is shorter than the period of those waves in the beam which also take part in scattering. If these are space-charge waves, the required criterion of dominant Compton scattering corresponds to the condition

$$\frac{\omega_1}{\sqrt{\gamma_0}} \frac{L/\gamma_0}{u_0} \ll 1 \tag{11.2}$$

where the first of the factors is responsible for the reduction in the Langmuir frequency in the frame of reference of the beam (see section 2.4) and the second, for the Lorentz contraction of the interaction length L in the beam reference frame; γ_0 is the relativistic factor (2.29). Criterion (11.2) is easily rewritten in terms of the current of the strongly relativistic beam

$$I < I_B \left(\frac{k_z}{k_\perp}\right)^2 \left(\frac{\gamma_0}{\pi}\right)^2 \tag{11.3}$$

where $k_z = \pi/L$, $k_\perp$ is determined by relation (2.14) and I_B is the Bursian current given by formula (2.27). It is not difficult to see that the limiting current of single-particle scattering (on the right-hand side of (11.3)) is much less than the Bursian current in actual devices in which $k_z^2 \ll k_\perp^2$.

The Raman scattering regime dominates if

$$\frac{\Delta u'}{\omega_1'} < \lambda' \tag{11.4}$$

where $\Delta u'$, ω_1' and λ' are the velocity spread in the beam, the beam's Langmuir frequency and the wavelength of the scattered wave (all these quantities are given in the beam reference frame: $\omega_1' = \omega_1/\sqrt{\gamma_0}$, $\lambda' = \lambda_0/\gamma_0$). Relation (11.4) signifies that the current of a homogeneous rod-shaped beam of radius a must exceed the threshold

$$I > I_c = \frac{ma^2}{4e} u_0^3 \gamma_0^3 \frac{(\Delta u_z'/u_0)^2}{\lambda_0^2} \tag{11.5}$$

† The Raman–Mandelshtam–Landsberg scattering.

or

$$I_c = I_B \frac{\gamma_0}{(\gamma_0^{2/3} - 1)^{3/2}} \frac{(\Delta u_z'/u_0)^2}{k_\perp^2 (\lambda')^2}. \tag{11.6}$$

In the case of a hollow beam (cross section radius a, beam thickness $\Delta \ll a$), we have

$$I_c = \frac{ma^2}{4e} \frac{2\Delta}{a} u_0^3 \gamma_0^3 \frac{(\Delta u_z'/u_0)^2}{\lambda_0^2}. \tag{11.5'}$$

In this last case, the Bursian current exceeds the limiting current of the homogeneous, rod-shaped beam. If the waveguide radius is much higher than the beam radius, the current excess is negligible but if the beam propagates close to a metal wall (i.e., if the gap separating the beam and the waveguide wall is less than one beam radius), the Bursian current is higher than the non-hollow beam current by a factor of a/Δ; for instance, it may be higher by two orders of magnitude than the limiting current of a beam which propagates at a considerable distance from the wall (see section 6.1). In the case of a hollow beam, therefore, the I_c/I_B ratio may be quite small. The ratio of I_c and I_B will be elaborated below, in specific examples of masers. It will also be clearly demonstrated that relations (11.3) and (11.4)–(11.6) are incompatible.

A survey of the principles implemented in new generator and amplifier designs will begin with a reminder of elementary processes which are at the basis of such phenomena as relativistic Doppler effect and the Compton (Thomson) scattering of electromagnetic waves by electrons. Correspondingly, let us look at the following model situations.

1. A source moving at a velocity u_0 relative to a laboratory frame of reference emits an electromagnetic wave of frequency ν' at an angle Θ' relative to the direction of velocity (the quantities ν' and Θ' are in the reference frame of the source). The frequency of the emitted radiation in the laboratory frame of reference is

$$\nu = \nu' \frac{1 + \beta \cos \Theta'}{(1 - \beta^2)^{1/2}} = \nu' \gamma_0 (1 + \beta \cos \Theta'). \tag{11.7}$$

If the source is relativistic (i.e., $\beta \equiv u_0/c \to 1$), then the frequency of the forward emission ($\Theta' = 0$) is

$$\nu = \nu' \left(\frac{1 + \beta}{1 - \beta}\right)^{1/2} \simeq 2\gamma_0 \nu' \tag{11.8}$$

and that of the backward emission ($\Theta' = \pi$) is

$$\nu = \nu' \left(\frac{1 - \beta}{1 + \beta}\right)^{1/2} = \gamma_0 (1 - \beta)\nu'. \tag{11.9}$$

We see that, unlike in the 'ordinary' (first-order) Doppler effect (the forward emission frequency is higher and that of the backward emission is lower than ν'), the relativistic (second-order) Doppler effect increases the frequency of radiation emitted by a moving source by a factor of γ_0. For example, only the second, 'transverse' Doppler effect survives when the direction of motion is perpendicular to the velocity of the source.

2. An electromagnetic wave of frequency $\nu = \omega/2\pi$ with phase velocity $v_{ph} = \omega/k$ and wavelength $\lambda = 2\pi/k$ (all these quantities are considered in the laboratory frame of reference) is incident on an electron beam which travels at a velocity u_0. By virtue of (11.7), the wave frequency in the reference frame of the beam is

$$\nu' = \frac{\omega'}{2\pi} = \gamma_0 \nu \left(1 - \frac{u_0}{v_{ph}} \cos \Theta\right) = \gamma_0(\omega - \boldsymbol{k}\boldsymbol{u}_0). \tag{11.10}$$

If the wave propagates in the direction opposite to that of the beam propagation $(\cos \Theta = -1)$, then

$$\nu' = \gamma_0 \nu + \gamma_0 \frac{u_0}{\lambda}. \tag{11.11}$$

3. An electromagnetic wave of frequency ν_0 (in the laboratory frame of reference) undergoes Thomson scattering by the electron beam. The wave first 'goes over to the beam's reference frame' (causing corresponding electron oscillations); the accompanying conversion of oscillation frequency is described by relation (11.10). By virtue of (11.7), the frequency (in the laboratory reference frame) of the wave scattered by electrons is

$$\nu_s = \nu_0 \gamma_0 (1 + \beta \cos \Theta') \left(1 - \frac{u_0}{v_{ph}} \cos \Theta\right). \tag{11.12}$$

For instance, if the wave is incident at $u_0 \simeq c$ towards the beam $(\cos \Theta = -1)$ and is scattered in the beam propagation direction $(\cos \Theta' = 1)$, then

$$\nu_s = 4\gamma_0^2 \nu_0. \tag{11.13}$$

If the wave follows the beam direction but the scattering is in the opposite direction, then

$$\nu_s = \gamma_0^2 (1 - \beta)^2 \nu_0. \tag{11.14}$$

We see that if need be, the emission frequency can be not only raised but considerably lowered as well.

4. An electron beam passes through an undulator, that is, a system in which the direction of the transverse magnetic field alternates along the beam (in actual devices, this is either a chain of deflecting magnets with alternating polarity of the transverse magnetic field $H_\perp$, or a current

spiral; there may also be a longitudinal magnetic field but it is unimportant in this particular case). An undulator produces a spatial modulation of the transverse velocity of electrons with a period λ_0—see figure 11.2. In this situation the beam undergoes oscillations of its velocity in time, at frequency (11.11), so that for $\nu = 0$ we find

$$\nu' = \gamma_0 \frac{\beta c}{\lambda_0}. \tag{11.15}$$

In the laboratory frame of reference, the frequency of the wave that the beam 'scatters' (emits) is implied by (11.7):

$$\nu = \frac{\beta c}{\lambda_0} \gamma_0^2 (1 + \beta \cos \Theta'). \tag{11.16}$$

This gives

$$\nu \to 2\gamma_0^2 \frac{c}{\lambda_0} \tag{11.17}$$

if $\beta \to 1$ and $\Theta' \to 0$ (forward emission), and

$$\nu \to \gamma_0^2 \frac{c}{\lambda_0} (1 - \beta) \tag{11.18}$$

if $\Theta' \to \pi$ (backward emission).

Note that the relativistic transformations above, for example, (11.15), imply that the wave number in the beam reference frame is $k_0' = 2\pi/\lambda_0' = 2\gamma_0 k_0 = 4\pi\gamma_0/\lambda_0$.

Relations (11.7)–(11.18) show that the scattering of an electromagnetic wave by the electron beam makes possible a very substantial transformation of its frequency, both to increase and to reduce it. The use of the Doppler frequency conversion for electromagnetic wave generation was first suggested by Ginzburg [11.22]. To be more specific, I will always assume below that the frequency-increasing solution is selected.

For instance, relation (11.17) implies that an electron beam of 50 MeV energy ($\gamma_0 \simeq 10^2$) permits the transformation of centimetre waves ($\lambda_0 = 1$ cm) into visible light ($\lambda \simeq 5 \times 10^3$ Å). At beam electron energy of 25 MeV and wavelength $\lambda_0 = 3$ cm, the scattered radiation wavelength is already 6 μm (infrared range). In the case of a lower-energy beam (5 MeV) and at $\lambda_0 = 10$ cm, the minimum wavelength of scattered radiation is around 0.4 mm.

It is of interest to consider one more example related to the effect of atomic 'channeling', that is, the propagation of a beam of fast particles along atomic planes of a crystal [11.23]. In this case a particle is, from the standpoint of its transverse motion, in a potential well of the type

$$e\varphi(x) = e\varphi_0 \left(\frac{x}{a}\right)^2 \tag{11.19}$$

where $\varphi(x)$ is the potential, x is the transverse coordinate, e is the particle charge, $2a$ is the spacing between crystal planes and φ_0 is the typical potential (e.g., $2a \simeq 10^{-8}$ cm and $\varphi_0 = (20-30)$ eV). In this case the frequency of transverse oscillations of a particle in a crystal is

$$\nu_0 = \frac{1}{2\pi} \left(\frac{2e\varphi_0}{ma^2}\right)^{1/2} \tag{11.20}$$

where m is the particle mass. Since the electric field (and potential φ) grow proportionally to γ_0 in the transition to the beam's frame of reference, the beam modulation frequency in this reference frame is $\nu' = \gamma_0^{1/2}\nu_0$ and the frequency of forward radiation in the laboratory reference frame, as given by (11.8), is

$$\nu = 2\gamma_0\nu' = 2\gamma_0^{3/2}\nu_0. \tag{11.21}$$

For example, if the energy of channeled particles (electrons or positrons) is $W_1 \simeq 50$ MeV ($\gamma_0 \simeq 10^2$), the energy of emitted quanta is $h\nu \simeq 40$ keV (the range of hard x-ray radiation!).

It is important to remark that the Thomson-scattering mode we consider here (scattering without frequency being changed in the act itself) occurs when the photon frequency in the beam's reference frame is small in comparison with the electron rest energy,

$$\gamma_0 h\nu_0 \ll mc^2. \tag{11.22}$$

In the opposite case (which currently cannot be more than of purely theoretical interest), it is necessary to take into account the Compton energy exchange *per se* between a photon and an electron; in view of this exchange, the frequency of the scattered photon $\nu < 4\gamma_0^2\nu_0$, namely [11.24],

$$\nu = \frac{2\nu_0}{\frac{1}{2}\gamma_0^2 + 2\hbar\nu_0/(\gamma_0 mc^2)}.$$

We see that if $\gamma_0 h\nu_0 \gg mc^2$, then $h\nu \to \gamma_0 mc^2$: the maximum photon energy tends to that of the electron.

These examples show that it is possible, in principle, to generate (amplify) electromagnetic waves in ranges which are least accessible to other methods of wave generation, both classical and quantum.

We have discussed several examples of *spontaneous* scattering of waves by fast electron beams. The intensity of scattering increases greatly if it is *induced*, that is, if it is stimulated by the radiation quanta whose frequency equals the frequency of the scattered wave. This stimulated scattering of photons (of the standing electromagnetic wave) in an electron beam was first discussed by Kapitza and Dirac [11.25]. If the scattering is stimulated, one speaks of superradiance; if quanta are accumulated using mirrors, one speaks of the laser (maser) effect.

Some additional remarks will now be made on the relevant terminology. First, Compton scattering *per se* is usually interpreted as scattering accompanied with a change in momentum (wavelength) of the photon incident on the particle (electron); this effect holds also in the case when a particle is at rest, but if the particle is in motion, the Doppler effect is added. However, in all versions of FEL (with a possible exception of the above purely hypothetical channeling scenario), the pure Compton effect is vanishingly small and is neglected. It would be more correct, therefore, to speak of Thomson scattering, when the frequency is not changed in the act of scattering itself but the Doppler effect is superposed. On the other hand, if the theory of the Compton effect is applied to the scattering of a photon (or of an electromagnetic wave of arbitrary frequency) by a relativistic electron (i.e., if only the laws of conservation of momentum and energy are considered and the notion of the Doppler effect is ignored), we again obtain the correct result. It is in this sense that the term 'Compton scattering' can be used (and is indeed used).

The second remark refers to the term 'laser'. The initial message of this abbreviation (Light Amplification by Stimulated Emission of Radiation) invoked the amplification of light. In the case of microwaves (beginning with sub-millimetre waveband), the usual term is 'maser'. As for the generators (amplifiers) we were discussing in this chapter, that is, devices operating on 'free' electrons and enabling us to perform the Doppler conversion of the coherent emission of radiation in a very broad frequency range, the accepted term is 'laser'; its origin is disregarded.

11.3 Raman Scattering of Electromagnetic Waves by Negative-Energy Waves of Electron Beams (Theory)

In principle, the electron beam is an oscillatory system whose natural frequencies correspond to space-charge waves and cyclotron waves (in the latter case, a longitudinal magnetic field is assumed). In general, such a wave system is essentially non-linear. A non-linear coupling of waves is caused by the fact that two quantities in the equations of the electromagnetic field, the current density and the Lorentz force, are proportional to products of the variables, the particle density n, velocity v and magnetic field $H_\perp$, namely $j = env$ and $F = (e/c)[vH]$†. If the system is irradiated by three waves of different frequencies, which affect the variables n, v and H, then the expressions for j and F include terms which are proportional to products of the time-dependent fields

---

† The waves treated here are not purely electrostatic.

and correspond to oscillations at the combination (sum and difference) frequencies. Under favourable conditions, a parametric resonance builds up in the system and oscillations are efficiently pumped. The mechanism of parametric wave coupling can be clarified using as an example Raman scattering of the pumping wave by space charge waves and cyclotron beam waves in a laser (maser) based on an undulator (figure 11.2); the undulator produces a transverse magnetic field H_x whose sign alternates along the beam propagation direction. The spatial period of this field is $\lambda_0 = 2\pi/k_0$. Due to the field H_x in the laboratory reference frame, there is an alternating electric field in the beam's reference frame: $E'_y = \gamma_0(u_0/c)H_x$, which varies at a frequency $\omega'_0 = \gamma_0 k_0 u_0$† (the 'pumping' field). In actual conditions, this field is quite high: at $H_x = 700$ Oe, $u_0 \simeq c$, $\gamma_0 \simeq 5$ and we have $E'_y = 10^6 B$ cm^{-1}, which corresponds to an energy flux density in the pumping wave $c(E'_y)^2/8\pi \simeq 10^9$ W cm^{-2} (!). The pumping wave causes oscillations of electrons with very high oscillation velocities: $v'_{0y} = eE'_0/m\omega'_0 \simeq 0.15c$‡, which predetermines intense parametric processes. Indeed, in addition to the oscillations of velocity in the pumping wave, beam electrons undergo density oscillations in the space-charge wave in the beam at a frequency ω'_1, where $\omega'_1 = (4\pi n'_1 e^2/m)^{1/2}$ is the Langmuir frequency of the beam in its frame of reference (since $n'_1 = n_1/\gamma_0$, then $\omega'_1 = \omega_1/\gamma_0^{1/2}$, where ω_1 is the Langmuir frequency corresponding to the beam density n_1 in the laboratory reference frame). Coupling of these two waves causes an alternating current at the combination frequencies $\omega'_0 \pm \omega'_1$, with the plus sign corresponding to the fast wave carrying positive energy, and the minus sign corresponding to the slow space-charge wave carrying negative energy. In what follows, I invariably mean the slow wave since the coupling of waves is active and pumps the oscillations only if this wave takes part in the process (see below, and also chapter 4). The alternating current mentioned generates a new, scattered wave at the combination frequency $\omega_s = \omega'_0 - \omega'_1$. The scattered wave and the pumping wave, coupled by the Lorentz force, then generate a wave at a difference frequency equal to the frequency ω'_1 of the slow wave of the beam's space charge and thus enhance (stimulate) the process still further. After this stage, the above cycle of stimulated Raman scattering repeats itself. The process is of the type of instability, in which the slow negative-energy beam wave transfers energy to the scattered wave and thereby enhances itself (see chapter 4). The increment of this instability is

$$\gamma \simeq \frac{u_0}{c}(\omega_s\omega'_1)^{1/2}. \tag{11.23}$$

The reader shall recall that among the two space-charge waves, only the

---

† All quantities in the beam's reference frame are primed
‡ m invariably denotes the electron mass.

slow wave can take part in the interaction described: the beam does not have enough energy to excite the fast wave (which carries positive energy).

At the oscillation velocities evaluated above, and at realistic wave frequencies, the value of γ given by relation (11.23) corresponds to the characteristic length of the exponential growth of the oscillation amplitude which is of the order of only several centimetres. For the subsequent comparison of the theory with experimental data, I will give the formulas for the frequency of scattered radiation in the undulator set-up of figure 11.2. In the laboratory reference frame, this frequency is implied by formulas (11.16) and (11.17),

$$\omega_s = \gamma_0 \left(\gamma_0 \frac{2\pi u_0}{\lambda_0} - \omega_1' \right) (1 + \beta)$$

$$= \left(\frac{2\pi u_0}{\lambda_0} - \frac{\omega_1'}{\gamma_0} \right) (1 - \beta)^{-1}$$

which transforms, for the undulator scattering by the slow space-charge wave of the beam, to the expression

$$\omega_s = \left[2\pi u_0 / \lambda_0 - \omega_1/(\gamma_0^{3/2}) \right] (1 - \beta)^{-1}. \tag{11.24}$$

The longitudinal magnetic field in the example above was zero. If this field (H) is non-zero, the beam may evince a slow negative-energy cyclotron wave (see chapter 4); this wave may play a role just as active as that played by the slow space-charge wave. If we consider the case of undulator scattering by a slow cyclotron wave, the frequency of the scattered radiation in the laboratory reference frame is

$$\omega_s = \left[2\pi u_0 \lambda_0^{-1} - \Omega_H \gamma_0^{-1} \right] (1 - \beta)^{-1}. \tag{11.25}$$

This is the case of the so-called magneto-Raman scattering [11.26]. Note that if the expressions for natural beam frequencies—cyclotron and the Langmuir—are written in the beam's frame of reference (viz. $\Omega_H = eH/mc$ and $\omega_1' = \omega_1/\gamma_0^{1/2}$, respectively, see section 2.4), then expressions (11.24) and (11.25) transform to a very symmetric form:

$$\omega_s = \begin{cases} 2\gamma_0^2 (k_0 u_0 - \dfrac{\omega_1'}{\gamma_0}) & \text{[I]} \\[2ex] 2\gamma_0^2 (k_0 u_0 - \dfrac{\Omega_H}{\gamma_0}) & \text{[II]} \end{cases} \tag{11.26}$$

where **I** indicates the slow space-charge wave and **II** indicates the slow cyclotron wave.

It is important to emphasize a factor of principal importance: the numerators of relations (11.24) and (11.25) contain minus signs which correspond to negative-energy waves (positive-energy waves would imply plus signs). This is of principal importance since it becomes possible to subject the concept of negative-energy waves to a quantitative experimental test. The experiment (see below) indeed supports this concept and also the conclusion of the theory that the frequency of the first of the waves (11.26) is independent of H while the frequency of the second one decreases with increasing H.

The interaction of waves in Raman scattering can be clarified using the dispersion diagram of figure 11.3. It plots the function $\omega(k)$ for three waves: the electromagnetic wave in the waveguide ($\omega_t^2 = \omega_L^2 + k^2 c^2$, where ω_L is the threshold frequency of the waveguide) and two space-charge waves, one fast ($\omega_f = u_0(k + k_0) + \omega_1/\gamma_0^{3/2}$) and one slow ($\omega_s = u_0(k + k_0) - \omega_1/\gamma_0^{3/2}$). The sum $k + k_0$ appears in the sum as an effective wave number because the additional wave number k_0 is 'imposed' on the system by the undulator. This factor is found to be decisive: without it, the frequency of the slow wave $\omega_s < kc$ is always lower than the electromagnetic wave frequency in the waveguide, so that there can be no interaction between such (non-synchronous) waves (as for the fast space-charge wave, it has already been mentioned that it cannot be excited by the beam). An increase in the wave number ($k \rightarrow (k + k_0)$), however, causes the slow-wave and the electromagnetic-wave branches to intersect, so that these waves become synchronous at the intersection point and do interact. This interaction (the active wave coupling) can be characterized as an instability which 'functions' as an effective generator of HF oscillations. A similar situation occurs in the case of the slow cyclotron wave whose frequency is $\omega = u_0(k + k_0) - \Omega_H/\gamma_0$.

The frequencies of scattered waves determined by the intersection of these branches with the waveguide mode are given by relations (11.26). As for the fast cyclotron beam wave, it cannot realize the *active coupling* with the electromagnetic wave in the waveguide—for the same reason that the fast space-charge wave cannot do it; this means that the fast cyclotron beam wave cannot be used to create a generator.

The analysis given above assumed a situation in which the beam electrons—in the non-perturbed state—have *only the longitudinal* velocity component. In this situation, Raman-scattering FEL and FEM operate on negative-energy waves. There is, however, another type of FEM, in which the *transverse* component of non-perturbed velocity is sufficiently large: for example, $u_{\perp 0}/c \simeq 0.4$. Such a device can also operate using the *fast* cyclotron wave which carries *positive* energy; if $u_{\perp 0}/c$ is sufficiently large, this wave is unstable (owing to the relativistic nature of the rotational motion of electrons in the beam). In the limiting case ($u_{\perp 0} > u_{\parallel 0}$), devices that operate using the instability of the fast cyclotron wave of

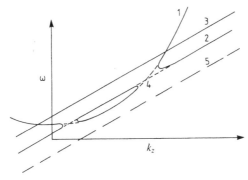

Figure 11.3 Dispersion diagram which illustrates the relationship between the space-charge wave, 2, and the electromagnetic wave in the waveguide, 1. Instability occurring in the vicinity of the intersection, 4, is used for the functioning of the Raman-scattering FEL. The equations of the branches are: 1 $\omega^2 = \omega_L^2 + k_z^2 c^2$; 2 $\omega = u_0(k_z + k_0) - \omega_1/\gamma_0^{3/2}$; 3 $\omega = u_0(k_z + k_0) + \omega_1/\gamma_0^{3/2}$. The broken curve, 5, traces the wave, 2, for $k_0 = 0$ (without undulator); it never intersects the wave, 1.

the rotating beam are known as cyclotron resonance masers (CRM), or *gyrotrons* [11.27]. Despite their advantages and a wide range of applications, they have a shortcoming: their oscillation frequency $\omega = \Omega_H/\gamma_0$ decreases with increasing energy of beam electrons. This shortcoming has been eliminated in devices which *combine* high transverse electron energy with still higher longitudinal energy. The relativistic behaviour caused by high *longitudinal* energy of beam electrons makes it possible to perform the Doppler enhancement of the emission frequency:

$$\omega = 2\gamma_\parallel^2\left(k_0 u_0 + \frac{\Omega_H}{\gamma_0}\right) \qquad (11.27)$$

(this expression is qualitatively different from the lower formula in (11.26) only in the sign in parentheses, which corresponds to the transition from the slow negative-energy wave to the fast positive-energy wave). In (11.27), $\gamma_\parallel$ is the relativistic factor which corresponds to the parallel energy of electrons. Now the emission frequency *increases* as the relativistic factor $\gamma_\parallel$ becomes higher; in principle, this facilitates the problem of advancing to the generation of still shorter waves.

The dispersion diagram for FEM of this type is shown in figure 11.4. The intersection of the waveguide-mode branch ($\omega_t^2 = \omega_L^2 + k_z^2 c^2$) with the fast cyclotron wave

$$\omega = (k_z + k_0)u_0 + \frac{\Omega_H}{\gamma_0} \qquad (11.28)$$

occurs at two points. The first of these (at low k_z) corresponds to the travelling-wave gyrotron, and the second (at high k_z) corresponds to the

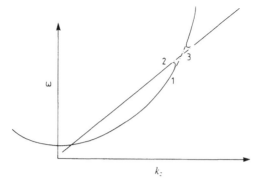

Figure 11.4   The dispersion diagram of the coupling of
the fast cyclotron wave, 2; with the electromagnetic wave in
the waveguide, 1. The instability which takes place at the
intersection (*3*) of the branches is used in the operation of the
cyclotron autoresonance maser (CARM).

FEM of the type we discuss here. In this last case, the frequency of the wave
generated is, under nearly resonant conditions $(k_0 u_0 \gamma_0 \simeq \Omega_H)$, around
$4\gamma_\parallel \Omega_H$. Generators of this type are known as cyclotron autoresonance
masers (CARM) [11.20, 11.28, 11.29]. The physical meaning of the term
autoresonance is that the beam slow-down (a drop in γ_0 caused by the
emission of waves) does not disrupt the resonance since the drop in u_0
in (11.28) is compensated for by an increase in Ω_H/γ_0, so that the wave
frequency remains almost unaltered [11.30]. Relation (11.28) shows that
the autoresonance takes place only for the positive-energy wave (the plus
sign on the right-hand side of (11.28)).

Generators of this sort are quite promising for the generation of
millimetre-band waves where their efficiency may reach 35 to 40%.

Undulator-based FEL have revealed the following feature: since γ_0
decreases as the beam slows down (as a result of energy loss to wave
excitation), the undulator period λ_0 along the beam propagation direction
must be gradually reduced so as to maintain the resonance conditions;
this is indeed done in the highest-efficiency FEL. This measure also helps to
push up the non-linear restriction on the FEL output, caused by the capture
of beam electrons into the potential well of the growing-amplitude wave.
The capture is faster, the steeper the beam is slowed down as its velocity
approaches the phase velocity of the wave; in the capture conditions, we
have

$$\frac{m}{2}\left(u - \frac{\omega}{k_z}\right)^2 \leqslant e\varphi_0$$

where φ_0 is the potential of the wave. An increase in k_z along the beam
(created by the variable-period undulator) compensates for the decrease in
u. If the beam electrons are captured and if the electric field $E_z \simeq k_z\varphi_0$ in
the potential well of the wave is sufficiently high, the oscillation frequency

of a captured particle in the well may exceed the instability increment; the instability is then disrupted, which sets the upper bound on the FEL output power†. (Of course, the capture violates the conditions of the hydrodynamic regime, with its small thermal velocity spread in the beam (see figure 4.2).) Calculations based on this mechanism of restriction of the FEL power output demonstrate that the restriction of generation occurs at a level at which up to 30% of the entire beam energy is converted into radiation [11.31].

An obvious shqrtcoming of the FEL with undulator is connected with the practical restriction of the undulator period from below. It was therefore proposed [11.32] to use as the pumping wave an oblique Langmuir wave propagating in the plasma. A similar idea was advanced in [11.33] where another possibility was also formulated in addition to the plasma wave: a pumping wave as a transverse electric or magnetic high-frequency field ($f_0 \gtrsim 10$ GHz, field amplitude $E \gtrsim 200$ kV cm^{-1}, the suggested wavelength band—submillimetres). The laser effect is connected with the interaction of the pumping wave with the negative-energy slow space-charge wave of the beam. It was also suggested, in the same framework, to construct an 'undulator' using a hypothetical structure with a sufficiently short spatial period, in a specially created laser plasma [11.34]. The authors believed that it would be possible to generate oscillations with 10 Å wavelength using a 150 MeV electron beam, and to generate γ radiation using electron beams of $\simeq 1$ GeV.

As any laser in general, an FEL uses mirrors, designed as a ring, through which the electron beam passes. If the amplification coefficient is very high, it becomes possible to completely eliminate mirrors from the system.

11.4   Free Electron Lasers (Masers) Using Compton and Raman Scattering (Experiment)

The possibility of designing—in principle—a laser based on induced Thomson scattering of electromagnetic waves by a relativistic electron beam‡ was first demonstrated in [11.35]. A 24 MeV electron beam ($\gamma_0 = 49$) with current up to 70 mA was sent along the axis of an undulator of the type shown in figure 11.2, with the longitudinal component of the magnetic field being 1 kOe and the transverse component, 2.4 kOe. The period of variation of the transverse field was $\lambda_0 = 3.2$ cm, the undulator length was 520 cm. The set-up included mirrors spaced by 12.7 m. A light beam of a CO_2 laser was sent along the electron beam in the opposite direction; its wavelength (10.6 μm) fits quite well relation (11.17), that is, $\lambda = \lambda_0/2\gamma_0^2$. It

---

† When the beam is captured by the wave, the beam instability (at a given frequency) is disrupted for the same reason which causes stabilization of the Landau damping [1.15].

‡ This section treats pulsed beams with pulse length of the order of tens of nanoseconds.

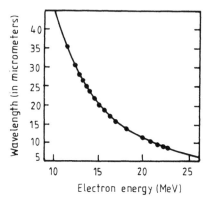

Figure 11.5 Wavelength generated by an induced Compton-scattering FEL as a function of electron energy of the relativistic beam [11.37].

was found that the system performed as a very good amplifier at precisely this wavelength; the principle it uses is thus induced Thomson (Compton) scattering.

The first experimental implementation of a free electron laser was reported by the same group of authors [11.36] for the electron current increased to several amperes; the electron energy was 43 MeV ($\gamma_0 = 87$). It was found that when the beam current exceeded a certain threshold (about 2 A), the emission line narrowed down dramatically, while the emission power jumped up by a factor of 10^8; this was obviously a laser effect. Good tunability of the laser wavelength was also demonstrated: $\lambda = 10.6$ μm at the beam electron energy $W_1 = 24$ MeV and $\lambda = 3.4$ μm at $W_1 = 43$ MeV (in accordance with formula (11.17)). The laser efficiency (the ratio of emission energy to beam energy) was rather low: about $10^{-2}\%$.

The essential advantages of the FEL as a generator tunable over a very wide range of frequencies were demonstrated most conclusively in [11.37]. The results of these experiments are shown in figure 11.5 which plots the emission wavelength as a function of beam electron energy: as electron energy grows from 11.5 to 23 MeV, the wavelength of the radiation generated decreases from 35 to 9 μm.

A physically similar result was obtained in [11.38] in the range of considerably longer wavelengths. In contrast to [11.37], the FEL operated in this case not on Compton but on Raman scattering (see figure 11.6). This study also revealed the dependence of power generated on the undulator field strength (see figure 11.7).

The FEL mentioned above (except the device in [11.38]) operated under the conditions of Compton (Thomson) scattering: the electron beam density was still too low. The power output was correspondingly low and the efficiency was $10^{-2}\%$. A novel result of principal importance was obtained by using a high-intensity beam in an undulator: beam electron energy

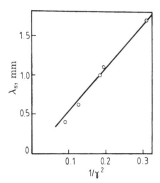

Figure 11.6 Wavelength generated by an induced Raman-scattering FEL as a function of electron energy of the relativistic beam. The undulator period: 0.83 cm [11.38].

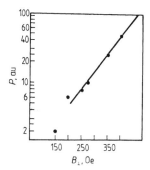

Figure 11.7 Power output of an Raman-scattering FEL as a function of magnetic field in the undulator [11.38].

1.2 MeV ($\gamma_0 = 3.4$), current 25 kA [11.39]. The undulator period was 0.8 cm, the generated wavelength $\lambda \approx \lambda_0/2\gamma_0^2 \simeq 0.4$ mm, the undulator field was $H_\perp = 400$ Oe; since the spatial increment of oscillations $1/\Gamma$ was of the order of several centimetres, we find that $\Gamma L \gg 1$. The power generated was about 1 MW, at 4% efficiency. In this experiment, the Debye radius of the beam was much smaller than the wavelength: condition (11.5) was satisfied. Indeed, since the radius of the (hollow) beam was 2.25 cm, its thickness was $\Delta \simeq 1.5$ mm, $\gamma = 3.4$ and the relative velocity spread was around 4%, we find, using (11.5′), that $I_c \simeq 200$ A. The beam current was therefore two orders of magnitude greater than the threshold I_c (it is rather difficult to estimate the factor by which the beam current was lower than the Bursian current since the width of the beam-to-wall gap was not mentioned in [11.39]).

The authors emphasized that their laser operated in the regime of collective (coherent) Raman scattering, using active coupling of the scattered wave to the slow negative-energy space-charge wave.

In a similar experiment of the same group of authors [11.40], a collective 'magneto-Raman' laser based on active coupling of the scattered wave to the slow cyclotron wave was realized. Under the conditions of this experiment ($W_1 = 0.75$ MeV, beam current $I = 5$–10 kA, $\lambda_0 = 0.6$ cm), criterion (11.5) was also satisfied. The radiation frequency approximately satisfied relations (11.25) and (11.26). The power output reached 10 MW at $\lambda = 1$ mm wavelength.

The physical mechanism of operation of the 'collective-Raman' FEL was studied in detail in [11.41, 11.42] where the spectral properties of stimulated Raman scattering were investigated. This experiment, which also involved an undulator, tested theoretical relations (11.24) and (11.25) for the parametric coupling of the scattered wave with negative-energy beam waves (for this reason, these relations had the minus sign in the numerators). The parameters of the system corresponded to regime (11.5), namely: $W_1 = 600$ keV, $I = 3$–10 kA, the beam density $n_1 = 3 \times 10^{11}$ cm^{-3} ($\omega_1 \approx 3 \times 10^{10}$ s^{-1}), the undulator magnetic field $H_\perp = 400$ Oe, the undulator period $\lambda_0 = 0.8$ cm.

Indeed, since the beam radius in this particular case was 1 cm, the thickness was about 1.5 mm and the velocity spread in the beam (a relatively large one) was about 7%, equation (11.5$'$) implies that for $\gamma_0 = 2.4$ the critical current was $I_c \simeq 500$ A. The beam current thus exceeded the threshold I_c by approximately an order of magnitude.

The longitudinal magnetic field was $H_0 = 12$ kOe, the oscillatory velocity of beam electrons was $u'_\sim \simeq 0.1\ c$, which is sufficient for the possibility of intense parametric processes. The experiments revealed two oscillation modes, one of which was identified unambiguously as related to the slow space-charge wave and the other, to the slow cyclotron wave. It was also shown quite clearly that as the magnetic field is changed, the frequency of the scattered wave in the regime of its coupling to the slow cyclotron wave varies in complete agreement with formula (11.25); when the electron density and energy are varied, the frequency of the mode which is determined by the space-charge wave varies in good agreement with formula (11.24). The working wavelengths were in the range from 1 to 3 mm. Although the FEL efficiency was not high in this experiment, it was shown nevertheless that the regime was still quite far from saturation: the emission power rose exponentially as the transverse component of magnetic field was increased, and the characteristic length of the spatial oscillation increment was several centimetres. This experiment has thus conclusively demonstrated that a laser based on collective Raman scattering implements the principle of active coupling of the scattered wave to the negative-energy waves of the electron beam.

The emission power output was substantially increased (at an impressively high efficiency) in [11.43], by reducing the beam velocity spread down to 10^{-1} %. The following regime was used: $W_1 = 1.35$ MeV,

$I = 1.5$ kA, $l_0 = 3$ cm, the longitudinal magnetic field $H_0 = 20$ kOe, the undulator field $H_\perp = 4$ kOe (greater by an order of magnitude than in [11.36]), the undulator length 63 cm, the emission frequency in the beam's frame of reference was close to the cyclotron frequency Ω_H, and in the laboratory reference frame it was about $2\gamma_0 \Omega_H$. The emission power output at the $\lambda = 4$ mm wavelength (which corresponded to relation (11.17)) was 35 MW, the efficiency reached 2.5% and the characteristic scale of the spatial oscillation increment was 4 cm. It is interesting that in this regime, which was still quite far from saturation, the efficiency achieved was only an order of magnitude below the theoretical limit (which comes to 20–30%). Since the beam velocity spread was very small, criterion (11.5) was satisfied with an especially large safety margin.

Implementations of CARM-type masers were described in a number of papers. For instance, the maser described in [11.44] had an undulator $l_0 = 2$ cm long, operated with a 350–600 keV electron beam at 0.4–1 kA current; it generated on the 4.3 mm wavelength and produced 6 MW power. In a slightly later paper [11.45], a CARM maser with an undulator $l_0 = 1.25$ cm long worked on a 800 keV electron beam on the 1.5 mm wavelength and produced 15 MW output. The transverse velocity of beam electrons ($v_\perp$) was quite high in this device: $\beta_\perp \equiv v_\perp/c \simeq 0.4$. The oscillation frequency obeyed relation (11.27) and thus increased with increasing field H_0.

CARM-type masers can be significantly improved if a regime in which $\omega - k_z u \gtrsim \Omega_H/\gamma_0$ is used. In this regime, the refractive index for the waves under consideration was greater than unity, and the emitted radiation was focused by the electron beam. As a result, it was possible to improve the diffractional divergence of the laser beam and, therefore, considerably increase the efficiency of wave generation. This efficiency is enhanced even more if the magnetic field increases along the beam (by about one third). With these measures taken, an efficiency of 79% was achieved in [11.46].

A different approach proves efficient for focusing the beam of a Raman-scattering laser: one uses a beam whose current slightly exceeds the limiting vacuum current ($I \gtrsim mc^3/e \simeq 17$ kA); in this case, the beam becomes essentially hollow owing to the shut-off by its own space charge, and acts as a waveguide [11.47].

The use of pulsed electron beams with energy from 1 to 100 MeV (pulse duration up to several tens of nanoseconds) produced some other very impressive results. For instance, a 8.6 mm wavelength generator of 1.9 GW of output power and 40% efficiency was built on the ETA Livermore accelerator [11.48, 11.49]. Such high power and efficiency were achieved mostly by reducing the undulator period and by considerably increasing the field strength along the undulator. This proved necessary because about 75% of beam electrons were found to be captured into the outgoing wave and thus to lose about 45% of their energy.

The collective Raman scattering FEL at 2.5 mm wavelength, producing

12 MW of power in 100 ns pulses at 7% efficiency is described in [11.50]. The beam electron energy was $W_1 = 750$ keV; the beam current, several kA; the beam diameter, 0.5 cm; the waveguide diameter, 0.63 cm; the longitudinal field $H_0 = 10$ kG; the undulator field $H_\perp = 10$ kG; the undulator length, $\simeq 40$ cm; its period, 1.45 cm; a 75% reduction of the undulator period along its length increased the efficiency by 3 to 12%.

Implementation of an efficient amplifier using an undulator with 1 MeV, 600 A electron beam was described in [11.51]. Its working frequency was 35 GHz, amplification coefficient—50 dB and output power—17 MW, at more than 3% efficiency.

The amplification coefficient of a Raman maser can be considerably improved by preliminary modulation of the electron beam by an electromagnetic wave [11.52].

An improvement of the maser's geometry [11.53] produced generator output of 250 MW at 10% efficiency (beam parameters: the beam electron energy 1 MeV, the current of several kA).

A survey of the (rapidly changing) current state-of-the-art in the field was given in [11.54, 11.55].

A number of experiments have demonstrated the feasibility of a two-stage FEL: they use, as an 'undulator' in the second stage, the wave which is generated by the electron beam when it passes through the undulator; after the first undulator, the beam enters the resonator which operates at the indicated wavelength. If the undulator period is λ_0, then the wavelength generated at the first stage is $\lambda_0/2\gamma_0^2$; at the second stage, it gets shorter by a factor of $4\gamma_0^2$ (see relation (11.13)); as a result, the wavelength of the generated radiation is $\lambda_0/8\gamma_0^4$. An example of such a generator is given in [11.56]: if the beam electron energy is 3 MeV (beam current from 2 to 20 A), the radiation wavelength is $\lambda = 1$ mm at the first stage and $\lambda = 5.3$ μm at the second; the output power after the second stage is up to 100 kW.

The highest-power visible-light generator was constructed [11.57] using an electron beam of $\simeq 100$ MeV. Its pulsed power output was 40 MW at 6000 Å wavelength. In other experiments [11.37], generation was at 10.6 μm wavelength and pulsed power of 10 MW was obtained, with beam electron energy ~ 20 MeV and beam current ~ 40 A.

Using a linear electron accelerator makes it possible to reach generator power $\simeq 400$ MW at 4% efficiency at the 10.6 μm wavelength [11.53].

Work has also started on using storage rings for visible light generation at electron energy of hundreds of MeV [11.58, 11.59].

Interesting ideas concerning the application of plasma waves for pumping Raman lasers were outlined in [11.60] (see also the references cited therein).

For a description of the current status of high-current relativistic-plasma HF electronics, the reader can address the papers [11.61, 11.62].

The papers [11.63, 11.64] analyse the feasibility of successful application of various FEL for heating plasma and for sustaining the electron current in closed systems such as tokamaks, which are being developed with a view to achieving controlled thermonuclear fusion.

The reader can obtain further details on FEL and FEM in the review [11.21] and also in the papers and conference proceedings cited therein.

A very original idea has been described recently: to realize a FEL by passing an intense relativistic electron beam $(18\,\text{MeV} \times 1\,\text{kA})$ through a high-density plasma $(n \simeq 4 \times 10^{15}\,\text{cm}^{-3})$ which is pre-modulated by short-wavelength ion sound (the idea of the so-called 'ion-ripple' laser). The authors of the publication hoped that this 'undulator' would be capable of generating ultraviolet laser radiation $(\lambda \simeq 1000\,\text{Å})$ with power output of tens of MW [11.65–11.67]. For some recent work in this field see [11.68, 11.69].

References

[11.1] Birdsall C K and Bridges W B 1963 *J. Appl. Phys.* **34** 2946

[11.2] Mahaffey R A, Sprangle P, Kapetanakos C A and Colden S 1977 *Phys. Rev. Lett.* **39** 843

[11.3] Kwan T J T and Thode L E 1984 *Phys. Fluids* **27** 1570

[11.4] Kwan T J T 1986 *Phys. Rev. Lett.* **57** 1895

[11.5] Peratt A L, Snell C M and Thode L E 1985 *IEEE Trans.* PS-**13** 498

[11.6] Davis H A, Bartsch R R, Thode L E, Sherwood E G and Stringfield R M 1985 *Phys. Rev. Lett.* **55** 2293

[11.7] Scarpetti R D and Burkhart S C 1985 *IEEE Trans.* PS-**13** 506

[11.8] Brandt H E 1985 *IEEE Trans.* PS-**13** 513

[11.9] Sze H, Benford J, Woo W and Harteneck R 1986 *Phys. Fluids* **29** 3873

[11.10] Woo W Y 1987 *Phys. Fluids* **30** 239

[11.11] Grigoryev V P and Zherlitsyn A G and Koval T V 1990 *Fizika Plasmy* **16** 1353 (Engl. Transl. 1990 *Sov. J. Plasma Phys.* **16** 784)

[11.12] Burkhart S 1987 *J. Appl. Phys.* **62** 75

[11.13] Davis H A, Bartsch R K, Kwan T J T, Sherwood E G and Stringfield R M 1988 *IEEE Trans.* PS-**16** 192

[11.14] Davis H A, Fulton R D, Sherwood E G and Kwan T J T 1990 *IEEE Trans.* PS-**18** 611

[11.15] Kwan T J T and Davis H A 1988 *IEEE Trans.* **16** 185

[11.16] Barkhausen V H and Kurz K 1920 *Phys. Zeits.* **21** 1

[11.17] Huttlin G A, Bushell M S, Conrad D B, Davis D P, Ebersole K L, Judy D C, Lezcano P A, Litz M S, Pereira N R, Ruth B G, Weidenheimer D M and Agec F J 1990 *IEEE Trans.* **18** 618

[11.18] Platt R, Anderson B, Christofferson J, Enns J, Haworth M, Metz J, Pelletier P, Rupp R and Voss D 1989 *Appl. Phys. Lett.* **54** 1215

[11.19] Marshall T C 1985 *Free Electron Lasers* (New York: Macmillan)

[11.20] Ginzburg N S and Petelin M I 1984 *Relativistic High-Frequency Electronics* No 4, ed A V Gaponov–Grekhov (Gorkii: AN SSSR) p 49

[11.21] Roberson C W and Sprangle P 1989 *Phys. Fluids* B **1** 3

[11.22] Ginzburg V L 1947 *Izv. AN SSSR Ser. fiz* **11** 165

[11.23] Kumakhov M A 1986 *Emission of Radiation by Channeled Particles in Crystals* (Moscow: Energoatomizdat)

[11.24] Arutyanian F R, Tumanian V A 1963 *Phys. Lett.* **4** 176

[11.25] Kapitza P L and Dirac P A M 1933 *Proc. Cambr. Phil. Soc.* **29** 297

[11.26] Granatstein V L, Schlesinger S P, Herndon M, Parker R K and Pasour J A 1977 *Appl. Phys. Lett.* **30** 384

[11.27] Gaponov A V, Petelin M I and Yulpatov V K 1967 *Izv. Vuzov Radiofizika* **10** 1414 (Engl. Transl. 1967 *Radiophys. Quantum Electron.* **10** 794)

[11.28] Petelin M I 1974 *Izv. Vuzov. Radiofizika* **17** 902; 1974 *Radiophys. Quantum Electronics* **17** 686

[11.29] Chu K R and Lin A T 1988 *IEEE Trans.* PS-16 90

[11.30] Kolomenskii A A and Lebedev A N 1962 *Dokl. AN SSSR* **145** 1259 (Engl. Transl. 1963 *Sov. Phys.–Doklady* **7** 745)

[11.31] Kolomenskii A A and Lebedev A N 1978 *Kvantovaya elektronika* **8** (Engl. Transl. 1978 *Sov. J. Quant. Electron.* **8** 879)

[11.32] Balakirev V A, Miroshnichenko V I and Fainberg Ya B 1986 *Fizika Plasmy* **12** 983 (Engl. Transl. 1986 *Sov. J. Plasma Phys.* **12** 563)

[11.33] Yan Y T and Dawson J M 1986 *Phys. Rev. Lett.* **57** 1599

[11.34] Loeb A and Eliezer S 1986 *Phys. Rev. Lett.* **56** 2252

[11.35] Elias R L, Fairbank W M, Madey J M J, Schwettman H A and Smith T J 1976 *Phys. Rev. Lett.* **36** 717

[11.36] Deacon D A G, Elias L R, Madey J M T, Ramian G I Schwettman H A and Smith T J 1977 *Phys. Rev. Lett.* **38** 892

[11.37] Newnam B E, Warren R W, Sheffield R L, Stein W E, Lynch M T, Fraser J S, Goldstein J C, Sollid J E, Swann T A, Watson J M and Brau C A 1985 *IEEE J. Quant. Electr.* QE-21 867

[11.38] Birkett D S, Marshall T S, Schlesinger S P and McDermott D B 1981 *IEEE J. Quant. Electr.* QE-17 1348

[11.39] McDermott D B, Marshall T C, Schlesinger S P, Parker R K and Granatstein V L 1978 *Phys. Rev. Lett.* **41** 1368

[11.40] McDermott D B and Marshall T C 1980 *Phys. Quantum Electron.* **7** 509

[11.41] Gildenbach R M, Marshall T C and Schlesinger S P 1979 *Phys. Fluids* **22** 971

[11.42]   Masud J, Marshall T C, Schlesinger S P, Yee F G 1986 *Phys. Rev. Lett.* **56** 1567

[11.43]   Parker R K, Jackson R H, Gold S H, Freund H P, Granatstein V L, Efthimion P C, Herndon M and Kinkhead A K 1982 *Phys. Rev. Lett.* **48** 238

[11.44]   Botvinnik I E, Bratman V L, Volkov A B, Ginzburg N S, Denisov G G, Kolchugin B D, Ofitserov M M and Petelin M I 1982 *Pis'ma v Zh. Eksp. Teor. Fiz.* **35** 418 (Engl. Transl. 1982 *JETP-Lett.* **35** 516)

[11.45]   Grossman A, Marshall T C and Schlesinger S P 1983 *Phys. Fluids* **26** 337; 1983 *IEEE J. Quant. Electr.* QE-**19** 334

[11.46]   Kleva R G and Levush B 1990 *Phys. Fluids* B **2** 185

[11.47]   Tripathi V K and Liu C S 1990 *Phys. Fluids* B **2** 1949

[11.48]   Orzechowski T J, Anderson B R, Clark J C, Pawley W M, Paul A C, Prosnitz D, Scharlemann E T, Yarema S M, Hopkins D B, Sessler A M and Wurtele J S 1986 *Phys. Rev. Lett.* **57** 2172

[11.49]   Robinson A R 1987 *Science* **235** 27

[11.50]   Yee F G, Marshall T C and Schlesinger S P 1988 *IEEE Trans.* PS-**16** 162

[11.51]   Gold S H, Hardesty D L, Kinkead A K, Barnett L R and Granatstein V L 1984 *Phys. Rev. Lett.* **52** 1218

[11.52]   Wurtele J S, Bekefi G, Chu R and Xu K 1990 *Phys. Fluids* B **2** 401

[11.53]   Plack W M, Gold S H, Fliflet A W, Kirkpatrick D A, Manheimer W M, Lee R C, Granatstein V L, Hardesty D L, Kinkead A K and Sucy M 1990 *Phys. Fluids* B**2** 193

[11.54]   Granatstein V L, Fliflet A W, Levush A B and Antonsen T M Jr 1989 *Comments Plasma Phys. Contr. Fusion* **12** 217

[11.55]   Kim K-Je and Sessler A 1990 *Science* **250** 88

[11.56]   Segall S B et al 1983 *Free Electron Generators of Coherent Radiation* **453** 178

[11.57]   Roberson C W and Sprangle P 1989 *Phys. Fluids* B **1** 36

[11.58]   Vinokurov N A and Skrinsky A N 1977 *Proc. Tenth. Intern. Conf. High Energy Charged Particle Accelerations* (Serpukhov) vol **2** p 454

[11.59]   Billardon M, Elleaume P, Ortega J M, Bazin C, Bergher M, Velghe M, Petroff Y, Deacon D A G, Robinson K E and Madey J M J 1983 *Phys. Rev. Lett.* **51** 1652

[11.60]   Balakirev V A, Miroshnichenko V I and Fainberg Ya B 1986 *Fizika Plasmy* **12** 983

[11.61]   Kuzelev M V, Rukhadze A A, Strelkov P S and Shkvarunets A G 1987 *Fizika Plasmy* **13** 1370 (Engl. Transl. 1987 *Sov. J. Plasma Phys.* **13** 793)

[11.62]   Carmel Yu, Minami K, Lou W, Kens R A, Destler W W,

Granatstein V L, Abe D K and Rotgers J 1990 *IEEE Trans* PS-**18** 497

[11.63]   Wagner C E, Boehmer H and Caponi M Z 1986 *Fusion Technology* **10** 1030

[11.64]   Cohen B I, Cohen R H, Nevins W M C and Rognlien T D 1991 *Rev. Mod. Phys.* **63** 949

[11.65]   Chen K R, Dawson J M, Lin A T and Katsouleas T 1991 *Phys. Fluids* B **3** 1270

[11.66]   Chen K R and Dawson J M *Phys. Rev. Lett.* 1992 **68** 29

[11.67]   Wilks S March 1992 *Physics World* 26

[11.68]   Hafizi B and Robertson C W 1992 *Phys. Rev. Lett.* **68** 3539 (see also the literature cited therein)

[11.69]   Sessler A M, Whittum D H and Yu L H 1992 *Phys Rev. Lett.* **68** 309

Conclusion

We have thus considered a range of collective phenomena in plasma physics, non-linear physics, physical electronics and radiophysics. This range appears to be fairly broad and at the same time sufficiently exhaustive. The material selected for this book belongs to the fundamentals of physics of beams in plasmas and as such, cannot become obsolete. For this reason, the unending flow of new data on topics discussed in this book (numerous papers published in journals and presented at conferences) need not frighten the interested reader: this entire 'avalanche' conveniently finds proper places 'on the shelves' and is readily understood at the current level of knowledge. I have gained certain experience in this matter, trying constantly to add fresh material to the text, up to the moment of sending the manuscript to the publisher. I hope therefore that the book displays the bulk of the essential information that can be found in the current journals by the moment of publication.

To conclude, I wish to point out the following important feature. The book is connected to a large extent to my own research, a situation which is obviously fraught with a bias. I made a special effort to eliminate this bias as much as possible. The reader will be the judge of my success in this matter. My most recent participation in the experiments described in the book involves the work on solitons (see chapter 10). To those interested in the latest results of this work (which proved even more successful than we expected), I can recommend another monograph, M V Nezlin and E N Snezhkin *Rossby Vortices and Spiral Structures*, to be published soon by Springer Verlag.

Index

Attractor stochastic, 28
Aurora borealis, 177

Barkhausen–Kurz oscillations, 251
Bohm's theorem, 173

Caviton, 198, 201, 205, 208

Debye radius, 1, 4, 134, 267
Doppler effect
  anomalous, 40, 56, 57, 60, 63
  normal, 56, 57
  relativistic, 255
  transverse, 256
Double layer, 172, 179, 181, 183
Dynamic pressure force, 190

Feigenbaum turbulence scenario, 28
Free electron laser, maser, 259, 266

Gyrotron, 263

Instabilities
  beam-drift, 41, 44, 117, 133, 176
  beam, hydrodynamic, 4, 40, 49,
    254
  Budker–Buneman, 31, 35, 45,
    123, 133, 156, 174
  Bursian, 4, 14, 15, 21, 108, 250
  diocotron, 47, 141
  electron–electron, 5, 124
  Kelvin–Helmholtz, 64

modulational, 5, 188, 192, 194,
  197, 208, 221, 236, 242, 244
Pierce, 4, 15, 16, 23, 109, 123,
  134, 174, 176
plasma stabilization of, 133, 135
resonant regime of, 124
slipping stream, 47

Landau damping, 64, 206, 223
  inverse, 61
  stabilization, 265
Langmuir paradox, 6, 49
Langmuir waves, plasmons, 198,
  270
  modulation instability of, 193,
    219, 236
  self-compression of, 208, 227, 242
Langmuir–Bohm conditions, 173,
  174
Lighthill criterion, 193, 236
Limiting current, 4, 12, 21, 108,
  109, 250

Maser
  cyclotron autoresonance, 264

Plasma beam, 5, 19

Scattering
  Compton (Thomson), 255, 256
  induced, 258

Raman–Mandelstam–Landsberg, 90
  collective, 267
  spontaneous, 258
Soliton, 188, 206, 220, 242
  envelope, 194, 196, 197, 198
  high-amplitude, 198, 225
  ion-acoustic, 237
  KdV-type, 237
  Langmuir, 5, 194, 196, 198, 205, 219, 227, 242,
  Langmuir, oblique, 237
  space-charge, 237
Space charge neutralization, 159
Space charge waves, 50, 253, 261, 265, 268
Spicons, 198, 206

Turbotron, 250
Turbulence
  Langmuir strong, 200, 244

Undulator, 256, 262, 265, 267, 270

Vavilov–Cerenkov effect, 56, 61, 65
  induced, 40
Vircator, 14, 250, 252
Virtual anode, 23
Virtual cathode, 12, 117, 146, 154, 174, 175, 251

Waves
  cnoidal, 205
  collapse of, 206, 209
  cyclotron slow, 253, 268
  Langmuir, 223, 226
  Langmuir, oblique, 228
  linear packet of, 198
  negative energy, 4, 65, 253, 260, 261, 262
  positive energy, 264
  space charge, 50, 253, 261, 265, 268
  Trivelpiece–Gould, 239, 242

Wave packet, 187

Milton Keynes UK
Ingram Content Group UK Ltd.
UKHW040445071024
449327UK00020B/1018